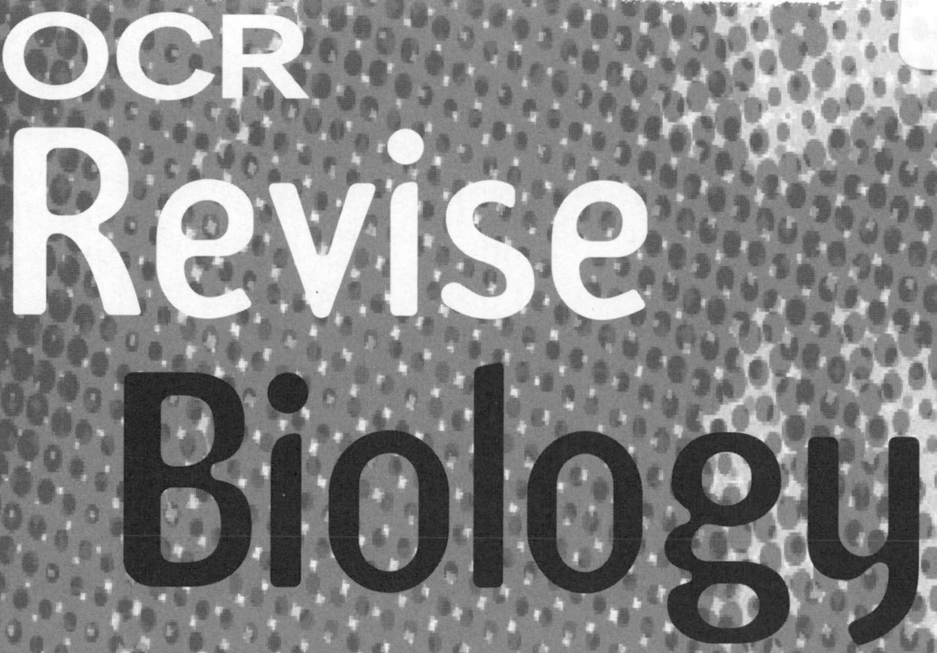

Exclusively endorsed by OCR for GCE Biology

Second Edition

Richard Fosbery, Ianto Stevens and Jennifer Gregory
Series editor: Sue Hocking

www.heinemann.co.uk

✓ Free online support
✓ Useful weblinks
✓ 24 hour online ordering

01865 888080

In Exclusive Partnership

Heinemann is an imprint of Pearson Education Limited, a company incorporated in England and Wales, having its registered office at Edinburgh Gate, Harlow, Essex, CM20 2JE. Registered company number: 872828

www.heinemann.co.uk

Heinemann is a registered trademark of Pearson Education Limited

Text © Richard Fosbery, Ianto Stevens, Jennifer Gregory 2008

First published 2001
This edition 2008

12 11 10 09 08
10 9 8 7 6 5 4 3 2

British Library Cataloguing in Publication Data is available from the British Library on request.

ISBN 978 0 435583 70 5

Copyright notice
All rights reserved. No part of this publication may be reproduced in any form or by any means (including photocopying or storing it in any medium by electronic means and whether or not transiently or incidentally to some other use of this publication) without the written permission of the copyright owner, except in accordance with the provisions of the Copyright, Designs and Patents Act 1988 or under the terms of a licence issued by the Copyright Licensing Agency, Saffron House, 6–10 Kirby Street, London EC1N 8TS (www.cla.co.uk). Applications for the copyright owner's written permission should be addressed to the publisher.

Edited by Anne Sweetmore
Designed by Wearset Ltd, Boldon, Tyne and Wear
Project managed and typeset by Wearset Ltd, Boldon, Tyne and Wear
Illustrated by Wearset Ltd, Boldon, Tyne and Wear
Cover photo of a mallow pollen grain © Alamy Images
Printed in the UK by Ashford Colour Press Ltd

Every effort has been made to contact copyright holders of material reproduced in this book. Any omissions will be rectified in subsequent printings if notice is given to the publishers.

Contents

Introduction

How to use this revision guide

This revision guide is for the OCR Biology AS course. Here is a plan of the units and modules you will study.

Unit	Module	Length of exam	Total number of marks in the exam (% of AS)
F211: Cells, exchange and transport	**Module 1: Cells**	1 hour	60 (30%)
	Module 2: Exchange and transport		
F212: Molecules, biodiversity, food and health	**Module 1: Biological molecules**	1 hour 45 minutes	100 (50%)
	Module 2: Food and health		
	Module 3: Biodiversity and evolution		
F213: Practical skills in biology 1	Practical tasks that will be set by OCR and marked by your teachers		40 (20%)

When you are revising, have a copy of the specification to hand. It is written as learning outcomes that tell you exactly what you should learn. As you revise, tick off each learning outcome when you understand what it means, and can write about and explain it.

This book follows the sequence of learning outcomes in the specification. Each module is divided into a number of spreads, which end with some **quick check questions** to test your recall and understanding. At the end of each unit there are **end-of-unit questions**, which resemble the types of question you will find in the examination papers. **Answers** to all the questions are on pages 90 to 98.

It is most important that you know the meanings of all the terms in the learning outcomes. These terms, and some others, are highlighted in bold in the text.

As part of your revision, you may wish to follow up some of the ideas in the course, especially in Unit 2 (F212). One good source of information – among many books and websites available – is www.askabiologist.org.uk.

We hope you enjoy your course and find this book useful. Good luck with your revision!

Unit 1 (F211) – Cells, exchange and transport

This unit is your introduction to AS Biology. It is a short unit of two modules that you may take in the January examination session after just 4 months.

Unit 2 (F212) – Molecules, biodiversity, food and health

This unit is longer than Unit F211, as it is designed to be taken in the summer examination session. The unit is divided into three modules.

Your practical work is assessed in Unit F213. You will need the theory from Units F211 and F212 to support your practical work. You can expect questions on practical work in the examination papers.

Topic (in this book)	Pages in this book	Learning outcomes from the specification	Useful ideas from GCSE
Unit 1 (F211), Module 1 – Cells, pages 2–17			
Microscopy	2–3	1.1.1 (a–d)	Using a light microscope to look at cells
Cell structure and function	4–5	1.1.1 (e–j)	Structure of plant and animal cells under the light microscope; parts of cells: mitochondria, chloroplasts, etc.
Cell membranes – structure and function	6–7	1.1.2 (a–e)	Much of this material is new, but you will know that some substances cross cell membranes and others do not; how drugs work
Transport across membranes	8–9	1.1.2 (f,g)	Diffusion, osmosis and active transport
Cell signalling	10–11	1.1.3 (a–i)	Hormones and chemical transmitters at synapses in the nervous system
Cell division and differentiation	12–15	1.1.2 (h–j)	Chromosomes, mitosis and meiosis; specialised cells
Tissues, organs and organ systems	16–17	1.1.3 (j–l)	Specialised cells; tissues and organs in plants and animals
Unit 1 (F211), Module 2 – Exchange and transport, pages 18–35			
Exchange surfaces and breathing	18–21	1.2.1 (a–i)	Lungs and breathing
Transport in animals	22–29	1.2.2 (a–n)	The heart, blood and blood vessels
Transport in plants	30–35	1.2.3 (a–m)	Transport of water in the xylem; transport of products of photosynthesis in the phloem
End-of-unit questions, pages 36–37			
Unit 2 (F212), Module 1 – Biological molecules, pages 38–59			
Water, proteins, carbohydrates and lipids	38–49	2.1.1 (a–s)	Proteins (enzymes); denaturing proteins; carbohydrates (starch and glucose); lipids, which you will know as fats and oils
Nucleic acids	50–53	2.1.2 (a–f)	DNA – the genetic material
Enzymes	54–59	2.1.3 (a–i)	Enzymes in digestion; the effects of factors such as temperature on the activity of enzymes; statins and controlling blood cholesterol
Unit 2 (F212), Module 2 – Food and health, pages 60–71			
Diet	60–61	2.2.1 (a–d)	Healthy and unhealthy diets; cholesterol and heart disease
Food production, food spoilage and food preservation	62–63	2.2.1 (e–j)	Food chains, pyramids of biomass; efficiency of food production; selective breeding; plant mineral nutrition; use of fertilisers
Health and disease	64–71	2.2.2 (a–r)	Disease; defence against disease; effects of smoking on the body; link between smoking and lung cancer
Unit 2 (F212), Module 3 – Biodiversity and evolution, pages 72–87			
Biodiversity	72–75	2.3.1 (a–h)	Ecology; continuous and discontinuous variation; adaptations of plants and animals
Classification and taxonomy	76–79	2.3.2 (a–h)	Much of this may be new to you
Evolution	80–83	2.3.3 (a–j)	Natural selection
Maintaining biodiversity – conservation	84–87	2.3.4 (a–g)	The effects of humans on the environment
End-of-unit questions, pages 88–89			

Studying and measuring cells

Key words

- resolution
- magnification
- staining
- light microscope
- electron microscope

Quick check 1

Examiner tip

When you do this sort of calculation, make sure you use the same unit for each measurement, e.g. millimetres (mm) or micrometres (μm). Remember there are 1000 μm in 1.0 mm.

Viewing cells

To see and study cells, biologists use various types of microscope. These produce a magnified image that can be drawn or photographed. Two key concepts in microscopy are **resolution** and **magnification**.

Resolution is the ability to see detail. An image of a cell is formed in a light microscope when light is focused; or in an electron microscope when electrons are focused. Particles or membranes in cells can be seen as separate objects if they are further apart than half the wavelength of light or of the beam of electrons used. Electron microscopes have much greater resolution because electron beams have a shorter wavelength than light.

Magnification is the ratio between the size of an object and its image. It is calculated using the formula:

$$\textbf{magnification} = \frac{\textbf{length of drawing or photograph}}{\textbf{length of object}}$$

You may be expected to calculate:

- the magnification of a photograph or a drawing, given the actual size in micrometres or given a scale bar
- the actual size of a cell or **organelle** from a photograph or a drawing, given its magnification.

The examples below should help you remember the steps to follow when doing these calculations.

Calculating magnification		Calculating actual size	
Diameter = 5 μm		Magnification = 18 500×	
Measure the diameter of the nucleus in millimetres	25 mm	Measure the length of the mitochondrion in millimetres	56 mm
Multiply by 1000 (μm)	25 000 μm	Multiply by 1000 (μm)	56 000 μm
Divide the diameter by the actual size	25 000/5 = 5 000	Divide the length in micrometres by the magnification	56 000/18 500 = 3.027
Round the answer up or down as appropriate		Round the answer up or down as appropriate	
Answer = 5000×		Answer = 3 μm	
You may be asked to express your answer to the nearest whole number.			

Quick check 2, 3

Examiner tip

When you make a drawing from the microscope, you magnify the image you see. Remember this when recording the magnification of your drawing.

Staining

To see cell structures you must add a stain, because most biological material is transparent. Some simple stains include:

- iodine solution, which stains **starch** grains black and other structures yellow
- methylene blue, which stains nuclei blue.

These stains give contrast to the material you are looking at. Biological material is also transparent to electrons. To give contrast in the transmission electron microscope (TEM), electron-dense materials are used as stains, for example the salts of the heavy metals lead and uranium. In the scanning electron microscope (SEM), heavy metals such as gold and palladium cover the object to be viewed. Electrons are scattered by the surface of the object, and this helps to create the image. In both cases the electrons strike a screen or are directed onto photographic film or onto a detector to make an image.

Examiner tip

You should be able to use your GCSE knowledge to name all the parts of animal and plant cells that you can see in a light microscope.

Light and electron microscopy

SEMs are useful for viewing the surfaces of samples. They produce an image like a photograph of a three-dimensional object. TEMs give an image of the internal structure of a thin section.

This table compares light and electron microscopes.

Feature	Light microscope (LM)	Electron microscope (EM)
Wavelength	Light, 400 nm	Electron beam, 1.0 nm
Resolution	200 nm	0.5 nm
Maximum useful magnification	×1500	×250 000 in TEM ×100 000 in SEM
Image	Natural colour (e.g. chlorophyll), Coloured if dyes or stains are used	Black and white – colour enhanced by computer to give false-colour images
Specimens	Living or non-living	Non-living
Advantages	Some living processes, such as mitosis, can be followed	Very high resolution – can see cell detail, e.g. cytoskeleton
Ease of use	Simple to use	Sample requires expert preparation

✓Quick check 4, 5

QUICK CHECK QUESTIONS

1. Explain why enlarging a photograph taken with a light microscope does not show more detail.
2. A plant cell is 40 µm in length. How long is its image when magnified ×3000?
3. An animal cell is 60 mm long when viewed ×4000. What is the cell's actual length?
4. Explain why samples of biological material are stained before viewing in the light and electron microscopes.
5. Explain the advantages of using the transmission and scanning electron microscopes to study cell structure.

Cells and organelles

Key words

- eukaryotic cells
- prokaryotic cells
- organelle
- ultrastructure
- cytoskeleton

Animal and plant cells

There is no such thing as a typical animal or plant cell. Cells are specialised to carry out certain functions, and as a result have special features. Two such cells are:

- a palisade mesophyll cell from a leaf that is specialised to carry out photosynthesis (see page 30)
- a pancreatic cell that secretes enzymes that work in the small intestine.

These are **eukaryotic cells**, as their genetic material is contained within nuclei. **Ultrastructure** is the fine detail of the **organelles** and **cytoskeleton** that can be seen in electron micrographs (EMs) taken with the TEM.

Examiner tip

Look at many EMs of different cells so that you learn to recognise these organelles.

Organelles

This table gives details about the functions of organelles in eukaryotic cells.

Organelle	Function(s)	Key points
Mitochondrion (plural mitochondria)	Aerobic respiration	Highly folded internal membrane to give large surface area for enzymes
Chloroplast	Photosynthesis	Grana made of stacks of membranes to give large surface area for chlorophyll and other pigments
Nucleus	Contains genetic information in DNA of chromosomes	Separated from cytoplasm by nuclear envelope with pores for communication between nucleus and cytoplasm
Nucleolus	Production of ribosomes	Dark-staining area in nucleus
Ribosome	Amino acids assembled to make proteins	On rough endoplasmic reticulum or free in cytoplasm
Rough endoplasmic reticulum (RER)	Site for ribosomes, transports proteins to Golgi apparatus	Outer surface covered in ribosomes
Smooth endoplasmic reticulum	Makes triglycerides (fats), phospholipids and cholesterol	No ribosomes on surface
Golgi apparatus	Modifies and packages proteins, makes secretory vesicles and lysosomes	Flat sacs of membrane formed from endoplasmic reticulum give rise to vesicles or lysosomes
Lysosome	Contains enzymes for destroying worn-out parts of cell and food particles	Membrane keeps enzymes separate from rest of cell
Centriole	Assembles microtubules for the cytoskeleton, and to form the spindle to move chromosomes when nuclei divide	Centrioles replicate before division so that they are at each pole of the cell (see page 13)
Cilium (plural cilia)	Moves liquids or materials along tubes (e.g. mucus along the trachea – see page 19)	Cilia and undulipodia have the same internal structure: a 9 + 2 arrangement of microtubules that slide over each other to cause cilia or undulipodia to move back and forth
Undulipodium (plural undulipodia)	Moves sperm cells and gametes of some plants (e.g. mosses and ferns)	

✓ *Quick check 1 and 2*

Microtubules and microfilaments make up the **cytoskeleton** that:

- supports animal cells (plant cells have cell walls for this purpose)
- moves organelles around eukaryotic cells
- moves cytoplasm, for example: during cell migrations in development; phagocytosis (page 66); cytokinesis (page 13).

Examiner tip

Note that there is **division of labour** within a eukaryotic cell as the organelles carry out different tasks for the cell.

Prokaryotes

Bacteria are **prokaryotes**. They have a much simpler cell structure than eukaryotes, and have no nucleus.

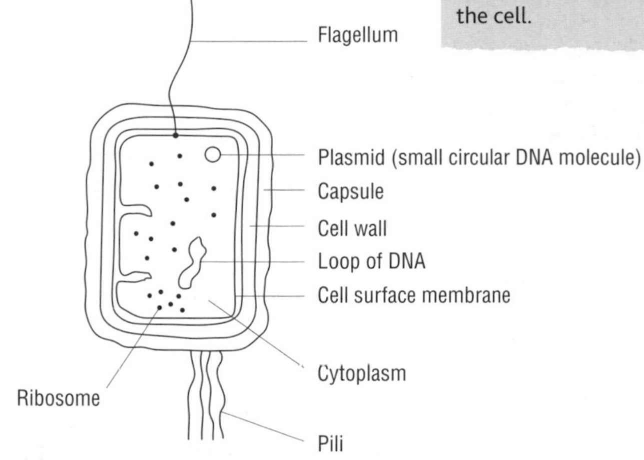

A prokaryotic cell

Prokaryotic and eukaryotic cells

This table compares the structure and size of prokaryotic and eukaryotic (animal and plant) cells.

Feature	Prokaryotic cells	Eukaryotic cells	
		Animal cell	Plant cell
Nucleus	✗	✓	✓
Cell wall	✓ (not cellulose)	✗	✓ (cellulose)
Mitochondria	✗	✓	✓
Chloroplasts	✗	✗	✓
Golgi apparatus	✗	✓	✓
Endoplasmic reticulum	✗	✓	✓
Vacuoles	✗	✓ (small, known as vesicles)	✓ (large – surrounded by tonoplast membrane)
Pili and flagella	✓ (in some)	✗	✗
Undulipodia and cilia	✗	✓ in some	✓ only in gametes of some plants
Typical diameter size/µm	0.5–3.0	10–20	40–100

Examiner tip

Include a column for features when making tables like this.

✓Quick check 3

QUICK CHECK QUESTIONS

1 Explain how division of labour occurs in plant and animal cells.

2 State the organelles from the table on page 4 that you would expect to find in: (i) a palisade cell; (ii) an enzyme-secreting pancreatic cell.

3 State four ways in which the structure of a eukaryotic cell differs from that of a prokaryotic cell.

Cell membranes

Key words

- fluid mosaic
- hydrophobic
- hydrophilic
- phospholipid bilayer
- partially permeable

Examiner tip

Most of the organelles you need to recognise are made of membrane – see page 4.

Hint

Phospholipids form a bilayer: 'heads out, tails in'.

Cell surface (plasma) membranes surround all cells and control what enters and leaves. Membranes divide up the cytoplasm of eukaryotic cells into separate compartments.

Membranes are very thin

Membranes are visible in the TEM at magnifications of ×100 000 as two dark lines separated by a clear space. The distance across the membrane is about 7 nm. Membranes are made of two layers of **phospholipid**, known as a bilayer, plus **protein**. The polar heads of the phospholipids are **hydrophilic** and are attracted to water. This is why they face towards the cytoplasm and towards the exterior of the cell. Both these areas are dominated by water. The hydrocarbon tails of the two layers are **hydrophobic** and are held together by weak hydrophobic bonds. Proteins are scattered about the membrane, and transmembrane proteins pass right through from one side to the other. **Carbohydrates** are attached to protein and **lipid**, and face the outside of the cell. This structure is called a **fluid mosaic**.

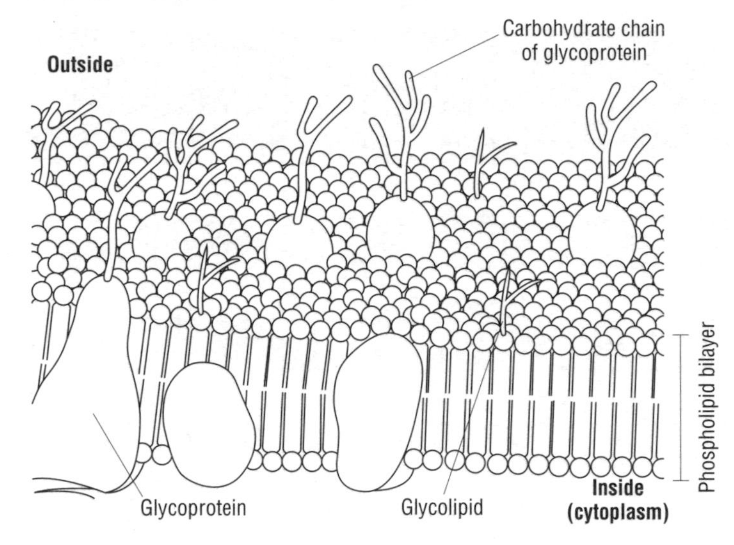

The phospholipid bilayer is split open to show the proteins that pass right through the membrane.

Quick check 1

Hint

Some proteins are anchored in membranes and do not move; others move like boats in a sea of phospholipid.

Examiner tip

You should be able to draw and label a diagram of a membrane like the one in the figure at the bottom of this page.

Why fluid mosaic?

- **Fluid**: phospholipids are liquid – think of a membrane as like a thin layer of oil.
- **Mosaic**: this is a picture made of many small pieces of tile. Proteins are like the pieces of tile surrounded by phospholipids, which are like the cement holding everything together.

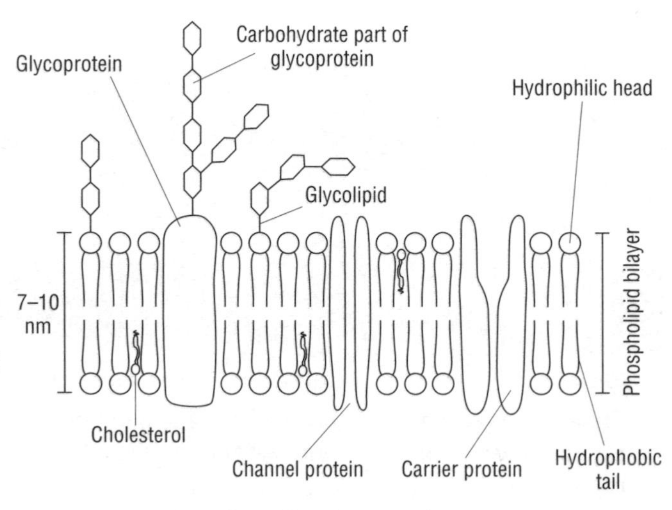

A cross-section through a cell surface membrane

Component of cell membrane	Functions
Phospholipids	• form a bilayer that acts as a barrier between cytoplasm and cell exterior • are fluid, so components can move within the membrane • are permeable to non-polar molecules such as oxygen and fatty acids • are permeable to small polar molecules such as ethanol, water and carbon dioxide • are impermeable to ions and large polar molecules such as sugars and amino acids
Cholesterol (only in eukaryotic cells – much more in animal cells than in plant cells)	• stabilises the phospholipid bilayer by binding to polar heads and non-polar tails of phospholipids • controls fluidity by preventing phospholipids solidifying at low temperatures and becoming too fluid at high temperatures • reduces permeability to water, ions and polar molecules
Proteins	• are transmembrane proteins acting as channels and carriers • are receptors for chemicals made by other cells, e.g. hormones
Glycolipids and glycoproteins (lipids and proteins with carbohydrate chains attached)	• are carbohydrate chains only found on the exterior surface of cell membranes • act as receptors for signalling molecules (e.g. hormones) and for drugs • act as cell surface markers that identify the cells to others (also known as cell surface antigens) • are involved in 'sticking' cells to one another (cell **adhesion**)

Functions of cell surface membranes

Quick check 2

Membranes are **partially permeable** because some substances pass through but others do not. The permeability of a membrane is determined by the phospholipids and proteins. Membranes are selective about what passes through. Cell surface membranes:

- act as a barrier to many water-soluble substances
- keep many large molecules, such as enzymes, within the cell
- are permeable to small molecules such as water, oxygen and carbon dioxide
- are permeable to selected molecules such as glucose and ions
- permit movement of substances by endocytosis and exocytosis
- permit recognition by other cells, such as those of the immune system
- provide receptors for signalling molecules such as hormones
- are often extended into microvilli to provide a large surface area for the absorption of substances by animal cells.

Examiner tip

Do not confuse microvilli with **cilia** – see pages 4 and 16.

Membranes within cells

The membranes within cells:

- divide the cell into compartments where functions can occur more efficiently
- isolate potentially harmful enzymes in lysosomes
- provide a large surface for pigments, such as chlorophyll, involved in photosynthesis in chloroplasts
- provide a large surface for holding the enzymes and coenzymes for forming ATP in mitochondria and chloroplasts
- surround vesicles that transport molecules between parts of the cell, e.g. that transport proteins from rough endoplasmic reticulum to Golgi apparatus.

Quick check 3

QUICK CHECK QUESTIONS

1 Explain why membranes are described as *fluid mosaic*.
2 Explain how phospholipids and proteins influence the permeability of a cell surface membrane.
3 Describe the distribution of membranes within animal and plant cells.

Hint

When answering quick check question 3, look back to the table of organelles on page 4.

Exchanges across membranes

Key words

- diffusion
- facilitated diffusion
- osmosis
- active transport
- bulk transport
- endocytosis
- exocytosis

Substances are exchanged between cells and their surroundings across cell surface membranes. Cells obtain all their requirements in this way, and their waste substances and some of their products pass in the opposite direction. Molecules move across membranes by:

- **diffusion**, including **facilitated diffusion** and **osmosis**
- **active transport**
- **bulk transport** (**endocytosis** and **exocytosis**).

Diffusion

Hint

See page 18 for more about diffusion in the **alveoli**.

Diffusion occurs passively – no ATP is required by the cell to move substances across the membrane because they move down their **concentration gradient**. Molecules such as oxygen and carbon dioxide diffuse across membranes as long as a concentration gradient exists. Both molecules are small and uncharged, so they diffuse through the phospholipid bilayer very easily.

Osmosis

Osmosis is a form of diffusion. Although water molecules are polar, they are small enough to pass through the phospholipid bilayer. However, the membranes of some cells (e.g. red blood cells) are very permeable to water as they have special, highly selective **channel proteins** for water, known as aquaporins. The movement of water into and out of cells is influenced by:

- the amount of water present in the cytoplasm and in the exterior environment
- the concentration of solutes, such as ions and sugars, on either side of the cell surface membrane
- the presence of aquaporins in membranes
- (in plants) the pressure exerted on cell contents by the cell wall, which is rigid and resists expansion and thus the uptake of water.

Examiner tip

Always explain the movement of water in and out of cells using **water potential** gradients. There are more examples on pages 30 and 36.

Water potential

Water potential is the tendency for water to move from one place to another, and is determined by the factors above. Solutions with a *high* water potential have *few* solute molecules; solutions with a *low* water potential have *many* dissolved solute molecules. Water moves from a solution with a high water potential to one with a lower water potential. The diffusion of water through a partially permeable membrane down a water potential gradient is osmosis.

Facilitated diffusion

Quick check 1

Many molecules that cells need are too large to pass between phospholipid molecules. They may also be charged and therefore unable to pass through the hydrophobic region in the centre of the bilayer. Protein molecules exist in membranes to help (or facilitate) the diffusion of these substances.

- **Channel proteins** are transmembrane proteins that form tunnels (or pores) through the bilayer for water-soluble molecules. Some channels are open all the time, others open when triggered by the presence of a signalling molecule such as a hormone. The lining of the pore allows water and polar substances to pass through.

- **Carrier proteins** change shape to help move molecules into or out of the cell. Molecules bind to the protein, which stimulates the protein to change its overall shape, so allowing the molecules to pass through the membrane.

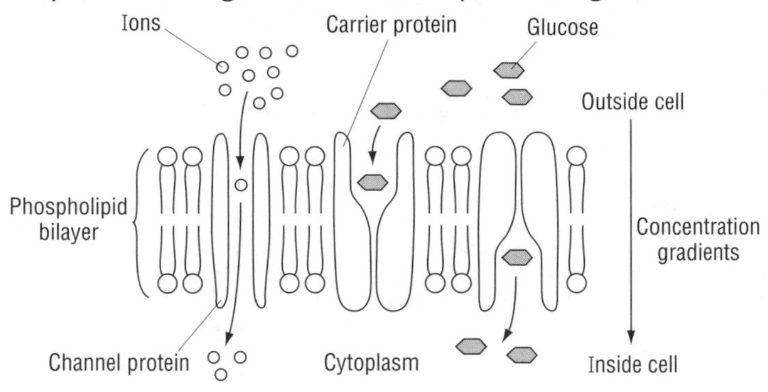

Facilitated diffusion through channel and carrier proteins

Active transport

Some substances required by cells are in a lower concentration outside the cell than inside it. Cells cannot obtain these substances by diffusion. Carrier proteins use energy from ATP made in **respiration** to move substances against their concentration gradient. **Root hair cells** absorb nutrients, such as potassium ions, from the water in the soil. Active transport is also used to pump molecules and ions out of cells.

Hint

Carrier proteins use energy from the cell in active transport.

Quick check 2 and 3

There are two forms of **bulk transport**.

- **Exocytosis**: substances packaged by the Golgi apparatus are delivered to the cell surface in vesicles, which fuse with the cell surface membrane to push out their contents.
- **Endocytosis**: some cells take up large molecules (such as proteins) and much larger solid objects (such as bacteria) by enclosing them inside vesicles or vacuoles formed by the cell surface membrane.

Examiner tip

The term 'nutrients' can be used to refer to many things – if you use it, always qualify it with an example, as here.

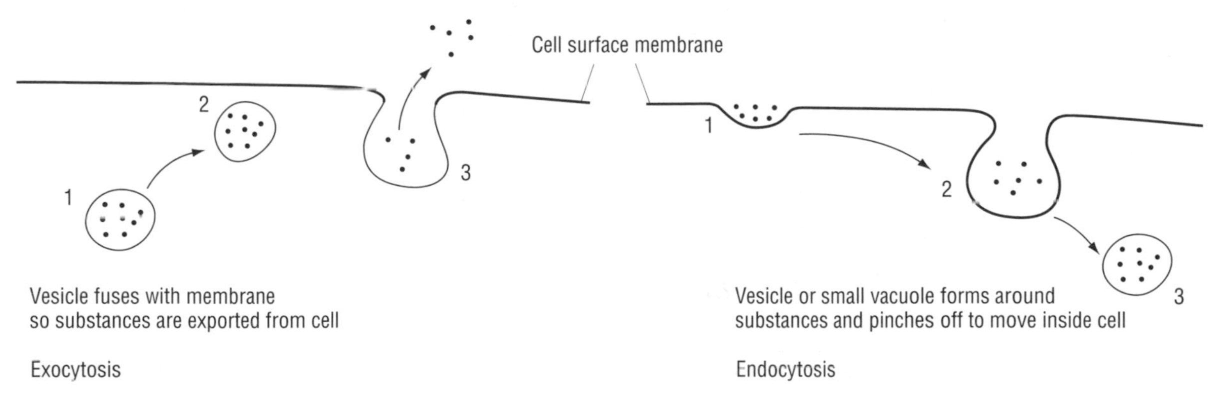

Cells use exocytosis to secrete substances and endocytosis to take in large molecules and particles

Quick check 4

QUICK CHECK QUESTIONS

1. Explain why carbon dioxide diffuses through phospholipid bilayers, but glucose does not.
2. State three ways in which active transport differs from facilitated diffusion.
3. Explain why root hair cells have mitochondria.
4. Define the following terms: diffusion; osmosis; facilitated diffusion; active transport; endocytosis; exocytosis.

Cell signalling and investigating cell membranes

Module 1

Cell signalling

Cells communicate with one another by **cell signalling**. There are three main ways in which animal cells can do this, as summarised in the diagram below.

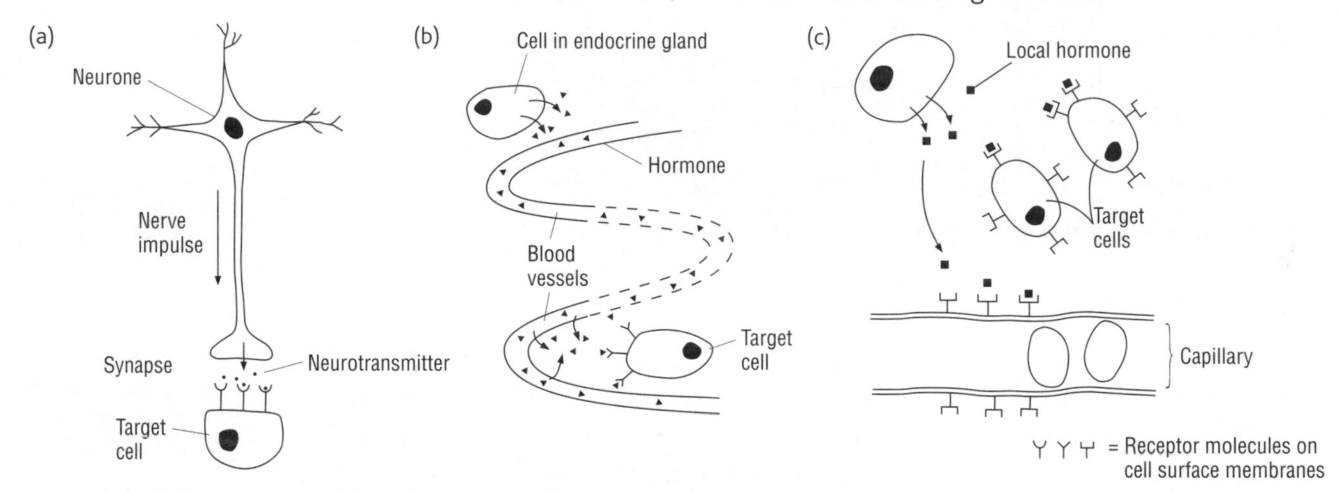

(a) Neurones send signals direct to target cells using neurotransmitters; (b) hormones travel long distances in the blood; (c) local hormones (such as histamine) stimulate cells in the immediate vicinity of the secretory cell

Key words

- neurone
- hormone
- receptor molecule
- water potential

Neurones

Neurones transmit information electrically over long distances. At the end of each neurone is a swelling that contains vesicles of a chemical transmitter substance known as a neurotransmitter. When a nerve impulse reaches the end of the neurone, these vesicles travel to the membrane and release molecules of neurotransmitter by exocytosis across the synapse. On the next neurone, or on an effector (such as a muscle cell or a gland cell), there are receptor molecules that detect the neurotransmitter molecules.

Examiner tip

Signalling molecules fit into their receptors like keys into locks. Their shapes are **complementary** – do not describe them as having the same shape.

Hormones

Endocrine cells secrete hormones into the bloodstream. Hormones have effects on their target cells, which are the only cells that detect the hormone and respond to it. **Receptor** molecules on the cell surface or inside the cell combine with the hormone, triggering a series of events inside the cell leading to the cell's response.

Examiner tip

Plants have signalling molecules too. These act like local hormones, but also travel long distances in the **xylem** and **phloem**.

Local hormones

Some cells release molecules that travel short distances to adjoining cells. Examples of these local hormones are histamine and cytokines. During an immune response, some lymphocytes release cytokines that stimulate other lymphocytes to divide by mitosis and secrete antibodies; cytokines also stimulate phagocytes to become more active (see page 66).

The hormones insulin and glucagon do not cross cell membranes as they are large, water-soluble molecules. Glycoprotein receptor molecules on the cell surface detect insulin.

Steroids (e.g. testosterone) are fat-soluble, so they can cross phospholipid bilayers and interact with receptors inside cells.

Quick check 1

Many drugs have a shape complementary to receptor molecules. **Agonists** act by mimicking the effect of the signalling molecule; for example, salbutamol (Ventolin) mimics adrenaline to relax **smooth muscle** in the bronchi, making it easier for asthmatics to breathe. **Antagonists** block receptors to stop the signalling molecule having any effect; for example, beta blockers block receptors for neurotransmitters in the heart and bring about a decrease in blood pressure.

Quick check 2

Examiner tip

As you read the descriptions below, remember that water diffuses *down* water potential gradients.

Investigating cell membranes

If you immerse animal and plant cells in solutions of different **water potential**, it is possible to see changes occurring if you look at the cells with a light microscope.

Cells immersed in:	Movement of water	Response of cells	
		Plant cells	Animal cells
Distilled water – higher water potential than cells	Water moves into cells by osmosis	Cells become turgid Cell wall prevents any more water entering	Cells swell Cells burst as there is no cell wall and the cell membrane is not strong enough to withstand the pressure
Dilute solution of salt or sugar – same water potential as cells	No net movement of water into or out of cells	Cells remain the same shape and volume	
Concentrated solution of salt or sugar – lower water potential than cells	Water moves out of cells by osmosis	Plasmolysis – vacuole shrinks, pulling cytoplasm and cell membrane away from cell wall Cells become plasmolysed	Cells decrease in volume and shrink

Betalain is the pigment that gives beetroot its dark red colour. If you soak pieces of beetroot in hot water, the pigment leaks out because the cell surface membranes and the vacuole membranes are damaged. This happens because:

- proteins are denatured by the heat and break up the structure of membranes
- phospholipids move more quickly at high temperature, making the membrane even more fluid and leading to its breakdown and the leakage of pigment.

Quick check 3

If you put beetroot tissue into water at different temperatures, you can find the lowest temperature at which the pigment first leaks out. You can use a colorimeter to record **quantitative** results from an investigation like this.

Quick check 4

QUICK CHECK QUESTIONS

1. Explain why cells need receptors to insulin, adrenaline and glucagon on their cell surface membranes, rather than inside the cytoplasm.
2. Explain how some drugs can mimic the effects of hormones and neurotransmitters or block their actions.
3. Explain why: (i) red blood cells burst when placed in distilled water but plant cells do not, and (ii) animal cells decrease in volume when placed into concentrated salt solutions.
4. Explain why betalain leaks out of beetroot tissue when it is placed into hot water.

Cell division and mitosis

Key words

- cell division
- mitosis
- stem cells
- meristematic cells
- chromosome

Examiner tip

Mitosis does *not* occur in prokaryotes. Yeast, like us, is a eukaryote.

Yeast is a single-celled fungus, which reproduces asexually by budding. As each cell grows, it produces a swelling (bud). The nucleus then divides into two so that the bud gains a nucleus. After a while the bud breaks off, leaving a bud scar. Sometimes buds start to produce their own buds before separating from the parent yeast cell.

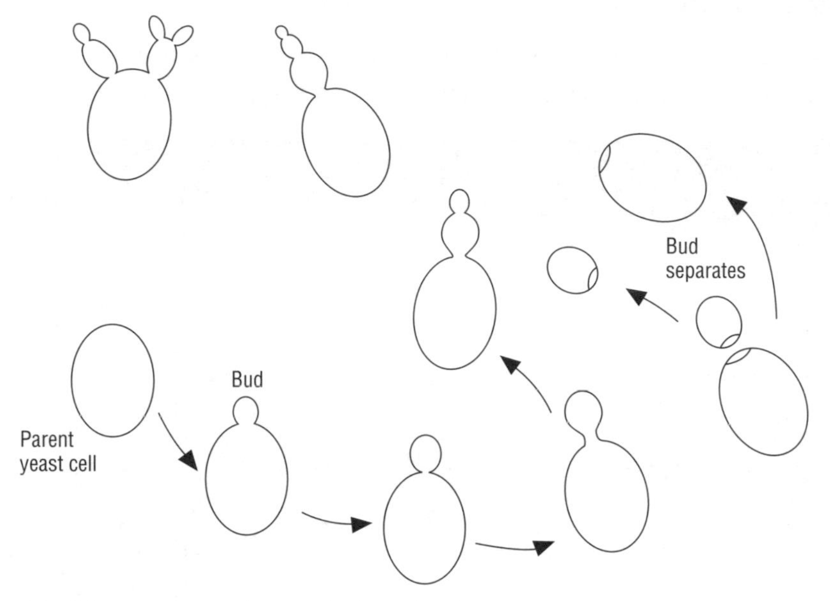

Cell division in yeast – the cells divide by budding

Unlike yeast cells, most cells grow to a certain size and then divide into two. The nucleus divides first, followed by the cytoplasm. The type of nuclear division that occurs in eukaryotes (such as yeast, other fungi, protoctists, plants and animals) is **mitosis**, which is involved in:

- growth, e.g. cells in an embryo divide to form tissue **stem cells**
- replacement of cells, e.g. tissue stem cells in bone marrow divide to produce red blood cells and neutrophils (see page 26)
- repair, e.g. in wound healing
- asexual reproduction.

Mitosis

The diagram opposite shows the mitotic cell cycle. A cell can divide only after it has replicated its **DNA** in the synthesis (S) stage that occurs during **interphase**. DNA synthesis and mitosis require energy. During the gap (G) stages, cells build up their energy reserves and make new membrane and organelles. As the DNA is replicated, the cell checks that the newly synthesised DNA is the same as the original DNA. During mitosis, the duplicated **chromosomes** separate and move to opposite ends of the cell. A cell may then divide into two to give two daughter cells. As a result of mitosis:

- the number of chromosomes in a nucleus stays the same
- the genetic information passed to the daughter cells is identical
- two new nuclei are formed
- no genetic variation occurs.

All cells in an organism have the same genetic information. If they did not, the organism would probably treat the cells as foreign and reject them.

Hint

Some cells have two or more nuclei because cell division has not followed nuclear division.

Stages of mitosis

1 **Prophase**: the DNA in chromosomes is packaged. Chromosomes shorten and thicken. This makes it easy for the cell to move chromosomes around. The chromosomes are now condensed and each chromosome has two **chromatids**. The **nuclear envelope** begins to break up into small pieces and disperse throughout the cell. **Centrioles** (not present in plant cells) move to opposite ends of the cell to make the poles.

2 **Metaphase**: chromosomes come to the middle of the cell. Centrioles (where present) organise microtubules into the **spindle** that stretches across the cell. Chromosomes are attached to the spindle at the **centromere**.

3 **Anaphase**: chromatids break apart at the centromere and are pulled by the spindle towards the poles. Once separated, the chromatids are chromosomes.

4 **Telophase**: nuclear envelopes re-form around each group of chromosomes at either end of the cell. The chromosomes uncoil.

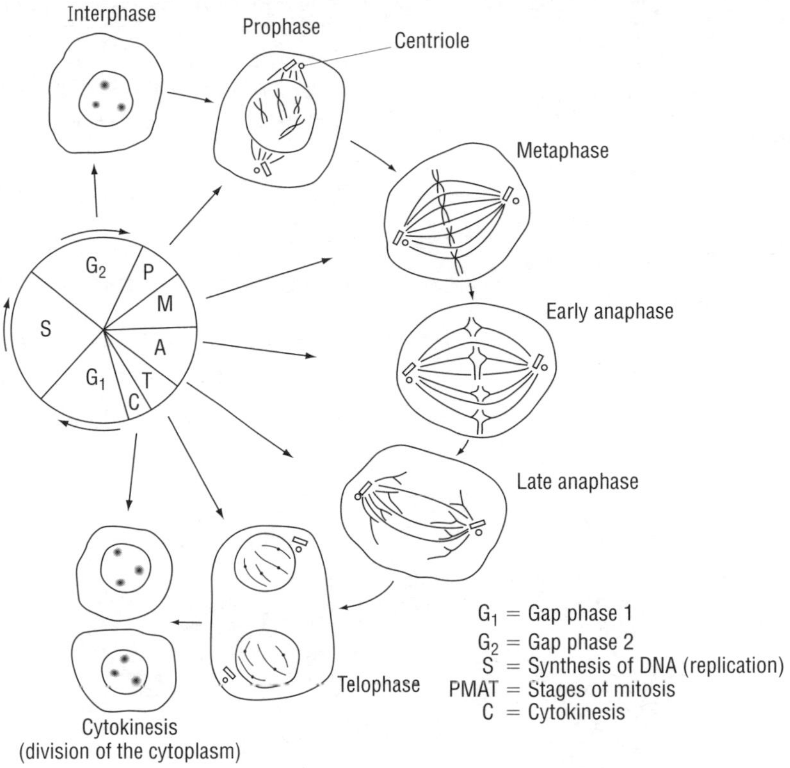

G_1 = Gap phase 1
G_2 = Gap phase 2
S = Synthesis of DNA (replication)
PMAT = Stages of mitosis
C = Cytokinesis

The mitotic cell cycle in an animal cell. Mitosis occupies 5–10% of the cell cycle. The length of time is exaggerated here

Hint

This must be an animal cell as it has centrioles and no cell wall.

Telophase is followed by an interphase in which DNA may be replicated if the cell is going to divide again.

During telophase, the cell may divide. In animal cells, microtubules form a 'draw string' just inside the membrane, which then fuses as it is 'pinched in'. In plant cells, microtubules direct vesicles to the middle of the cell to form a cell plate, which forms a new cell wall. New cell surface membrane is made on either side to enclose two separate cells. This stage of the cell cycle is **cytokinesis**. Note that yeast and some other eukaryotes have a closed mitosis in which the nuclear envelope does not break down and the spindle is formed *within* the nucleus.

✓*Quick check 1, 2, 3, 4*

Cells that have the ability to divide by mitosis and give rise to cells that change into specialised cells are called **stem cells** in animals, and **meristematic cells** in plants. The process of changing from a dividing cell into a specialised cell is called **differentiation**.

On page 27 you can read about stem cells in bone marrow differentiating into blood cells. Meristematic cells in plants differentiate into the cells of the xylem and phloem as well as all the other tissues in plants (see pages 30, 32 and 35).

QUICK CHECK QUESTIONS

1 Describe the behaviour of *one chromosome* during a mitotic cell cycle.
2 Explain why DNA replication must always occur before mitosis.
3 What happens to the quantity of DNA in a cell during: (i) replication; (ii) cytokinesis?
4 Stem cells divide repeatedly by mitosis. State what happens to the number of chromosomes in the nuclei of the daughter cells after each division.

Hint

In quick check question 1, the mitotic cell cycle includes interphase.

Chromosomes and meiosis

Module 1

Key words

- homologous chromosomes
- meiosis
- diploid
- haploid
- allele

✓Quick check 1

Chromosomes are made of DNA and a group of proteins known as **histones**. Proteins are in chromosomes to support and package the DNA, which is the genetic material.

Homologous chromosomes

The figure below (left) is based on a photograph of all the chromosomes from a human cell. On the right, the same chromosomes are arranged into homologous pairs.

When photographed like this, chromosomes are double structures. Replication has occurred, so each chromosome consists of two molecules of DNA that are packaged tightly to make two sister chromatids. As replication is very precise, the two sister chromatids are genetically identical.

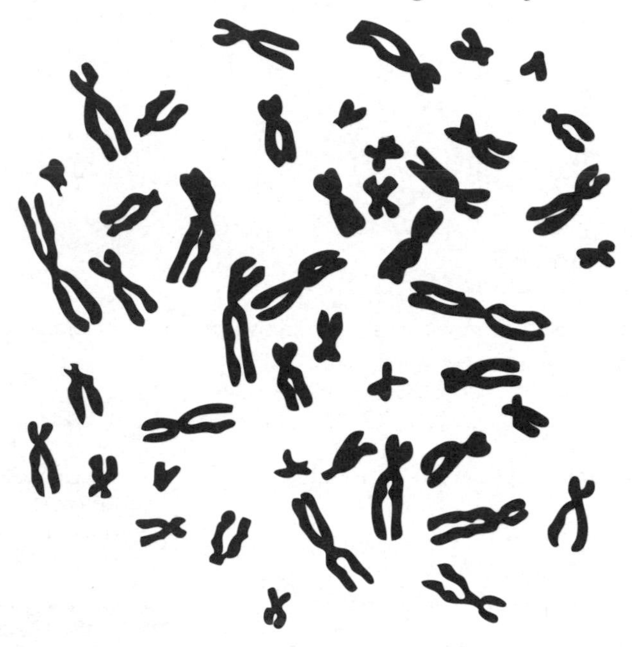

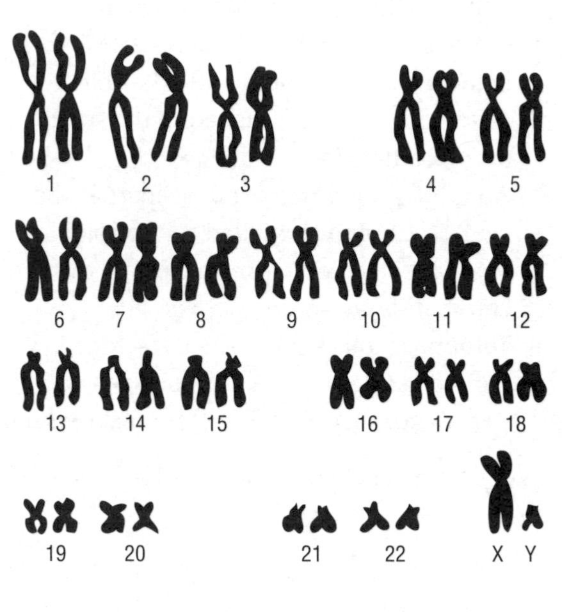

Hint

You inherited your chromosomes from your parents. In each **homologous** pair there is a maternal and a paternal chromosome.

Chromosomes from the figure on the left have been matched and put into pairs to give the arrangement on the right, which is called a karyotype. Pairs of chromosomes are homologous. The chromosomes in each pair are the same size, they have the same shape, the centromere is always in the same place, and they have the same **genes**. (Note that the sex chromosomes, X and Y, have only one small region that is homologous.)

Meiosis and life cycles

Examiner tip

In mitosis, the chromosome number stays constant; in meiosis it is halved.

The diploid number of humans is 46. Other species have different numbers – do not assume the diploid number of every species is 46.

From generation to generation, the chromosome number in the body cells of **species** that reproduce sexually remains constant. This number of chromosomes in the body cells is the **diploid** number. The human diploid number is 46. When **gametes** (eggs and sperm) are produced, a different type of nuclear division occurs – **meiosis**. This occurs only in special parts of the body, such as the testes and ovaries in animals, and the anthers and ovary in flowering plants.

During meiosis, the chromosome number is halved so human gametes have 23 chromosomes – one of each type. The number of chromosomes in gametes is the **haploid** number. For humans, the haploid number is 23. This reduction in chromosome number ensures that at fertilisation, when the gametes fuse to form a **zygote**, the diploid number of chromosomes is restored.

Meiosis involves two divisions of the nucleus, but it is during the first division that most of the variation is generated. Homologous chromosomes pair together during meiosis and then separate. This process ensures that each new nucleus receives one full complement of chromosomes – one chromosome from each pair.

Pairs of homologous chromosomes can be separated in a number of different ways. You can see in the figure below that with a diploid number of 4, there are 2^2 arrangements = 4. In a human cell with a diploid number of 46, there are $2^{23} = 8\,388\,608$ different ways in which homologous chromosomes can be arranged in pairs. This causes variation because, although a pair of homologous chromosomes carries the same genes, one or more of these genes may have variant forms (alleles; see page 82), so they control characteristics in different ways. You may know some examples of this from GCSE. For example, the gene that controls production of a protein carrier in cells lining the airways may not function correctly. Some people may inherit two copies of this faulty gene and develop cystic fibrosis.

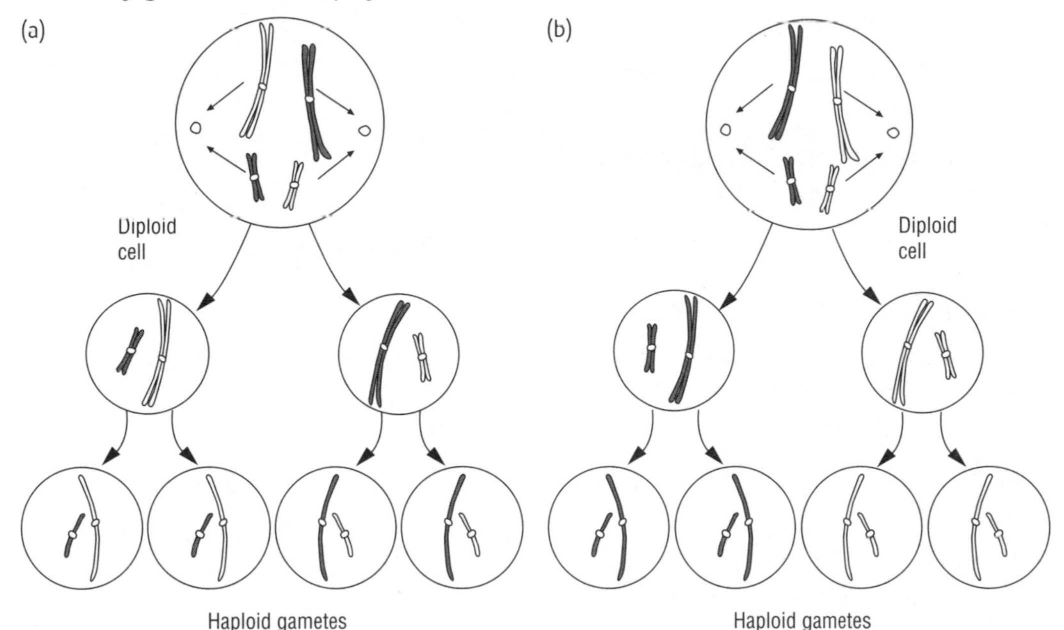

Two cells are dividing by meiosis: in (a) two pairs of chromosomes are arranged in one way; in (b) they are arranged differently. Notice how this leads to four different combinations of chromosomes in the gametes

Quick check 2, 3, 4

In animals, the haploid gametes are either spermatozoa (sperm) or ova (eggs). Human sperm are adapted for the role of delivering a haploid nucleus to an egg cell in the following ways.

- The undulipodium beats to move the sperm cell through the female reproductive tract.
- Mitochondria provide energy in the form of ATP for movement of the undulipodium.
- The acrosome at the tip of the sperm contains enzymes to digest a pathway through the cells and jelly coat that surround the egg.
- Sperm are compact and streamlined as they have little cytoplasm and a small nucleus with very highly condensed chromosomes.

QUICK CHECK QUESTIONS

1 Explain what is meant by *homologous pairs of chromosomes*.
2 Draw up a table to show the differences between mitosis and meiosis.
3 Explain why meiosis is described as a reduction division.
4 Explain the significance of meiosis in life cycles.

Hint

Use three columns when answering quick check question 2, and use the information on pages 12 and 13.

UNIT 1 Tissues, organs and organ systems

Module 1

Key words

- tissue
- organ
- organ system

Some small organisms, such as *Stentor*, have a body that is not divided into separate cells like ours. *Stentor* has cilia, which it uses for moving and feeding. The cilia beat together in a pattern that you can see here to create a current of water. *Stentor* uses its cilia to filter small particles of food from the water. Cilia are short, cylindrical projections from the cell surface that beat back and forth (see page 4).

- They are ideal for moving small organisms like *Stentor*.
- Animals use cilia for moving fluids past stationary cells.
- In the trachea, bronchi and bronchioles, cilia move a carpet of mucus.
- In fallopian tubes cilia move eggs from the ovary to the uterus.

Cilia

Food vacuole

Stentor, a freshwater protoctist with cilia that beat in a coordinated way to move it through water (×35)

Examiner tip

Do not confuse cilia with microvilli (see page 7). Cilia are larger and move backwards and forwards.

Animal tissues

In multicellular organisms, cells are not like *Stentor*: they are specialised to carry out specific functions and are often clustered together or in layers. In animals, a layer of ciliated cells forms a ciliated epithelium, which is a **tissue**.

✓ *Quick check 2*

A tissue is a group of similar, specialised cells in a multicellular organism that carries out a specific function, or several related functions. You can find examples of specialised cells throughout this guide – look for sperm cells, red blood cells, neutrophils, palisade cells, root hair cells and guard cells.

Hint

You have already read about sperm cells on page 15.

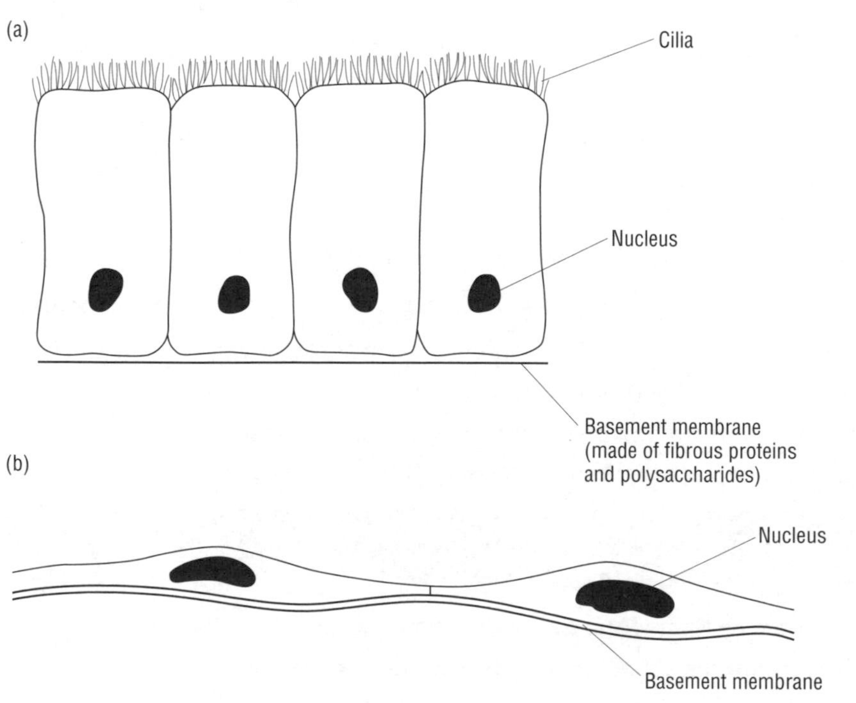

Two tissues in the lungs: (a) ciliated epithelium from the bronchus (see page 19); (b) squamous epithelium from the alveoli

The thin, flat cells that line the alveoli in the lungs form a tissue called squamous epithelium. Other examples of animal tissues are cartilage and bone (for support), muscle (for movement), and blood (for transport – see page 26).

Quick check 1

Tissues are grouped into **organs**. Animals have many organs, such as lungs, hearts and kidneys. Roots, stems and leaves are plant organs.

Quick check 2

An organ is a group of different tissues that form a distinct structure and function together.

Quick check 3

Plant tissues

Xylem and **phloem** are the transport tissues in plants. Xylem transports water and ions. Phloem transports sugars (mainly sucrose) and other compounds made by plants. The figure on the right shows where these tissues are located in a plant.

Sections cut through the leaf, stem and root of a plant to show the distribution of tissues. Plan drawings like these are used to show tissues – they do not show any cells

Organ systems

Organs work together to perform one major body function, such as excretion, digestion or coordination.

Organs including the trachea, lungs, diaphragm and rib cage work together to bring about ventilation of the gas exchange surface in the lungs. The heart and blood vessels work together to move blood around the body. See pages 18 and 24 for further details about these systems.

In multicellular organisms, cells are specialised to carry out specific tasks. There are some functions that all cells carry out – they all respire to provide themselves with energy; they all make their own proteins, membranes and organelles. But in an animal, cells deep in the body cannot obtain their own food and oxygen directly from the environment – they depend on cells in the gut and lungs to do this. They rely on the heart to keep a constant supply of blood providing the substances they need and removing their wastes. They also rely on the lungs and the kidneys to excrete this waste to the outside world.

Quick check 4, 5

QUICK CHECK QUESTIONS

1 Name four different tissues in animals, and state one function of each.
2 State the meanings of the terms *tissue* and *organ*.
3 Name two different plant tissues and state one function of each.
4 Explain the meaning of the term *division of labour*.
5 Explain, using examples from both plants and animals, how cells rely on each other for their survival.

Exchange surfaces and the lungs

Module 2

Key words

- surface-area-to-volume ratio
- diffusion gradient
- gas exchange
- alveolus

Hint

Remind yourself about the idea of *division of labour* in Module 1 – see pages 4–5 and 16–17.

Examiner tip

When you calculate SA:V ratios, always include ':1' or 'to 1' in your answer.

Quick check 1

Very small organisms, such as the protoctist *Stentor* (see page 16), do not have a specialised transport system. Oxygen from the surroundings diffuses through the cell surface membrane and the cytoplasm to mitochondria, where it is used in respiration. The distance from the edge of the cell to the centre is no more than 0.5 mm. Food is digested inside food vacuoles, and the digested food products diffuse from the vacuole to all parts of the cytoplasm. Large organisms cannot rely on diffusion from the surface or from their digestive systems for their supplies of oxygen and food. There are two reasons for this:

- the body surface is not large enough
- distances from the surface to the centre are too great.

Surface-area-to-volume ratios

Imagine that an organism has the shape of a cube. Also imagine that its body grows from a small cube with sides of 1 mm into a larger one with sides of 10 mm. Study the table carefully and you will see that as the organism grows, its volume (or mass) increases more quickly than its surface area. A large organism has less surface area per mm^3 of body than a small organism for each mm^3 of body. This means there is relatively less space on the body surface for uptake of oxygen and removal of carbon dioxide by diffusion. So, many organisms have special surfaces for gas exchange, such as gills or lungs.

Length of side/mm	Volume/mm^3	Surface area/mm^2	Surface-area-to-volume ratio (SA:V ratio)
1	1	6	6:1
5	125	150	1.2:1
10	1000	600	0.6:1

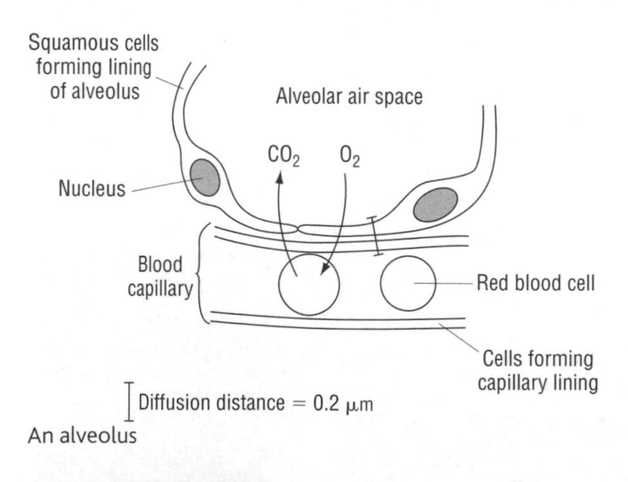

An alveolus

Few cells are really this large. We are using 'cells' of this size to show you the idea of the ratio.

Efficient exchange surfaces such as alveoli (left) have:

- a large surface area
- a thin layer (or layers) of cells that provide:
  - a short distance for diffusion
  - a good blood supply with many capillaries to absorb useful substances and/or to deliver waste substances.

The gas exchange surface in the lungs is made of many alveoli.

Quick check 2

Examiner tip

Exchange surfaces are also moist, but this is not a feature that makes them efficient.

In the lungs, oxygen diffuses from the air in the alveoli into the blood. Carbon dioxide diffuses in the reverse direction. This is gas exchange. To make this as efficient as possible, the air in the alveoli is continually refreshed so that the concentrations of oxygen and carbon dioxide are such that steep **diffusion gradients** are maintained between the air and the blood. Ventilation of the alveoli is achieved by moving the **diaphragm** and the **rib cage** when we breathe. Air moves through a system of airways to reach the alveoli.

Breathing mechanism

The movements of the diaphragm and the rib cage change the pressure in the lungs: when we breathe *in* the pressure is *less* than that of atmospheric air; when we breathe *out* the pressure is *greater* than that of atmospheric air. The difference in pressure between the air in the lungs and the atmosphere causes air to move in and out. The table summarises the movements and changes that occur.

	Breathing in (inspiration)	Breathing out (expiration)
Diaphragm	Contracts and moves down	Relaxes and is pushed up by pressure in the abdomen
Rib cage (ribs and intercostal muscles)	*External* intercostal muscles *contract* to move ribs upwards and outwards (*internal* intercostal muscles *relax*)	*External* intercostal muscles *relax* so that ribs fall with gravity (*internal* intercostal muscles *contract* during forced expiration)
Volume of thorax	Increases	Decreases
Pressure in lungs	Decreases (lower than atmospheric)	Increases (higher than atmospheric)

This table shows the distribution of the main tissues and cells in the airways.

Tissue/cell	Trachea	Bronchus	Bronchiole	Alveolus
Cartilage	✓	✓	✗	✗
Goblet cells	✓	✓	✓ (a few)	✗
Ciliated cells	✓	✓	✓	✗
Smooth muscle	✓	✓	✓	✗
Squamous epithelium	✗	✗	✗	✓
Elastic fibres	✓	✓	✓	✓

Hint

A diagram showing the rib cage and intercostal muscles is in Question 3 on page 37.

✓ Quick check 3

The following table shows the functions of cells, tissues and fibres in the gas exchange system.

Cartilage Provides strength to trachea and bronchus; holds open the airways so there is little resistance to air flow	**Smooth muscle** Contracts to narrow the airways
Goblet cells Secrete mucus Mucus is sticky and collects particles of dust, spores and bacteria	**Elastic fibres** Stretch when breathing in and filling the lungs Recoil when breathing out to help force air out of the lungs
Ciliated cells Move mucus up the airways towards the mouth	**Squamous epithelium** Gives a short diffusion pathway for oxygen and carbon dioxide in the alveoli

Examiner tip

Oxygen diffuses into red blood cells. The walls of both alveoli and capillaries are made of squamous epithelial cells; the diffusion distance is 0.2 µm.

QUICK CHECK QUESTIONS

1 Explain why humans have a specialised gas exchange surface, but protoctists do not.

2 How many cell surface membranes must oxygen diffuse through from the air in an alveolus to a haemoglobin molecule in a red blood cell?

3 Describe the pathway taken by air as it passes from the atmosphere to the alveoli.

Measuring lung activity

Key words

- spirometer
- tidal volume
- vital capacity
- oxygen consumption

Doctors measure lung activity to find out about people's lung capacity and how easy it is for them to breathe in and out. This is useful when checking the health of people with diseases of the lungs, such as chronic obstructive pulmonary disease (see page 68). Peak flow meters are used to detect the speed with which people can exhale the air from their lungs. This tells doctors how clear the airways are. A **spirometer** may be used to measure lung volumes and the rate at which oxygen is used. The figure above shows how a spirometer works.

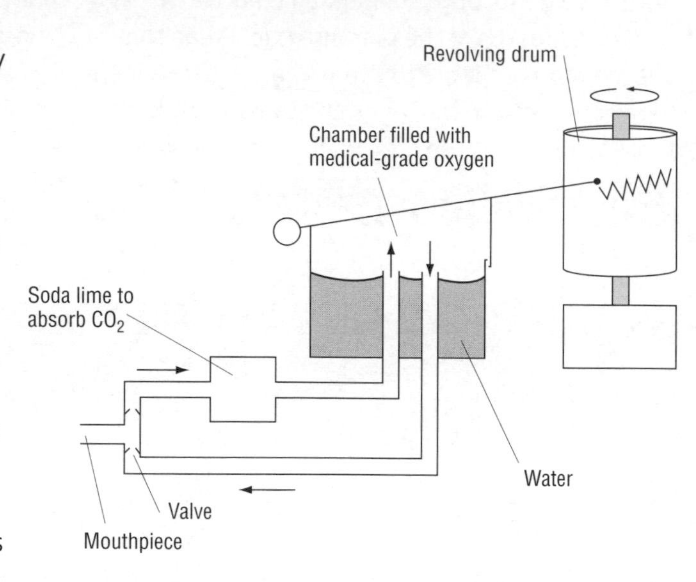

The following figure shows a spirometer trace of a 17-year-old male who breathed normally for 2 minutes, then took a deep breath and breathed out as much air from his lungs as possible.

- **Tidal volume** is the volume breathed into the lungs in one breath. At rest, it is usually 0.5 dm^3 (500 cm^3).
- **Vital capacity** is the maximum volume of air that can be breathed out of the lungs after taking a deep breath. This is often about 5 dm^3, but is dependent on many factors such as age, gender, fitness and state of health.

✓*Quick check 1*

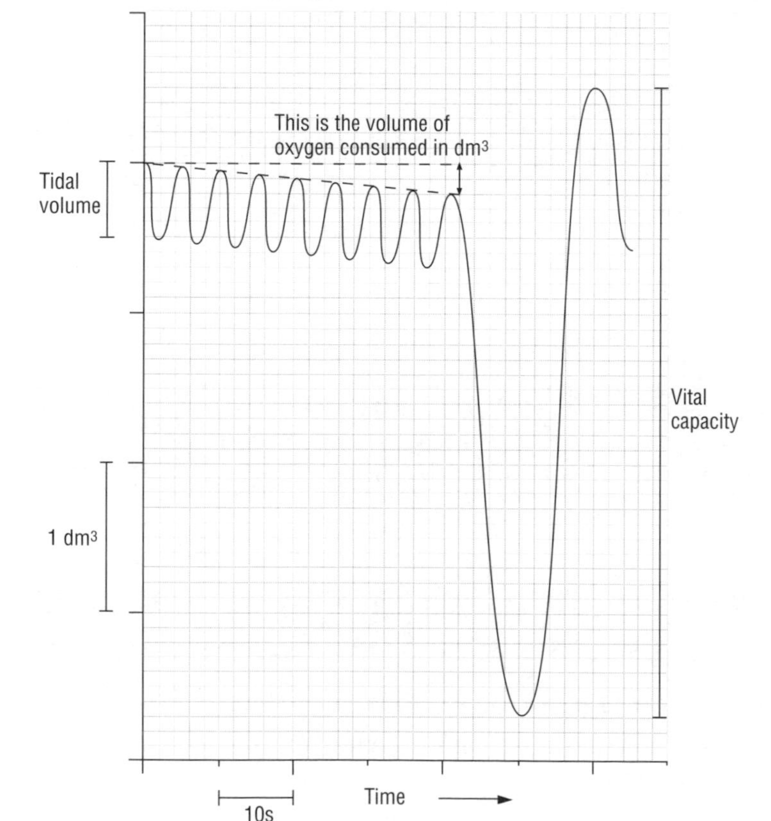

The next figure shows a spirometer trace of the same 17-year-old who breathed at rest for 1 minute.

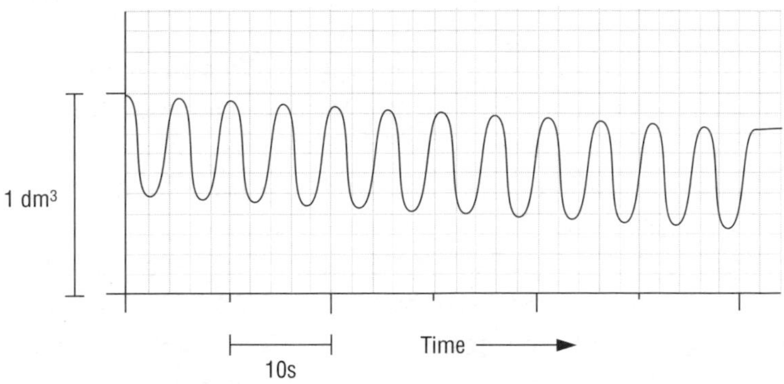

> **Examiner tip**
>
> When you take readings from a graph or from spirometer traces, use a ruler and pencil to make sure your readings are accurate.

He then took some vigorous exercise for a short period of time, and immediately he stopped exercising he made another trace which is shown in the following figure.

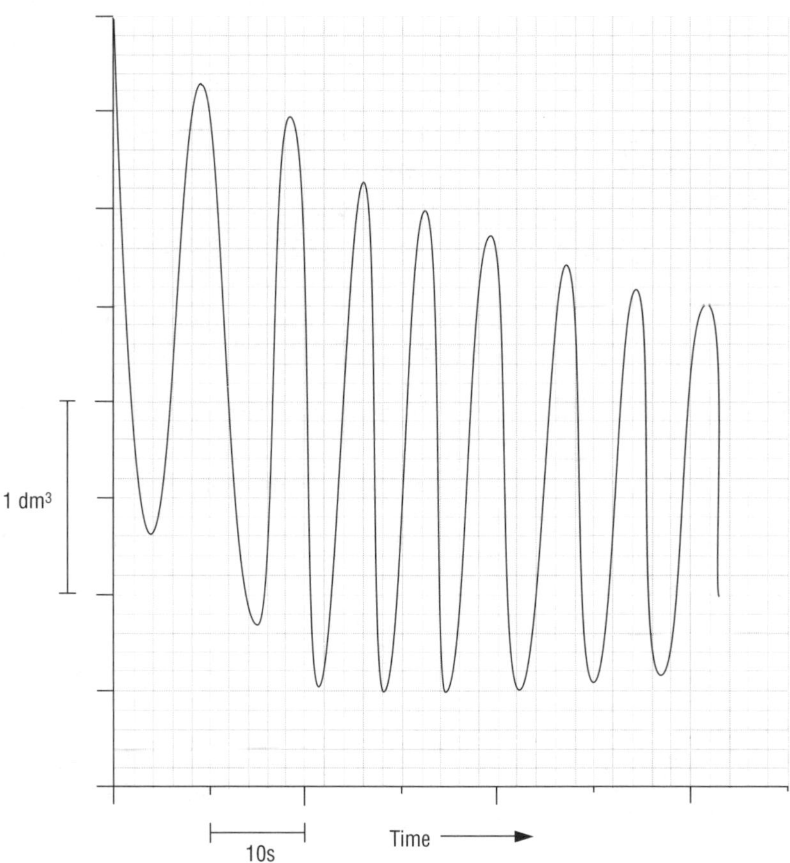

It is possible to calculate the following from the spirometer traces on this page:

- breathing rate/breaths per minute = number of peaks per minute
- mean tidal volume/dm^3 = mean of several tidal volumes from the trace
- ventilation rate/dm^3 per minute = volume of air taken into the lungs per minute = breathing rate × mean tidal volume
- oxygen consumption/dm^3 per minute = decrease in peaks over a period of 1 minute.

The figure on the opposite page shows how to take these measurements.

Quick check 2

QUICK CHECK QUESTIONS

1 Explain what is meant by the terms *tidal volume* and *vital capacity*.

2 Draw the table below and complete it to compare the breathing before and after exercise as shown in the three spirometer traces in this spread.

Feature	Before exercise	After exercise
Mean tidal volume/dm^3		
Vital capacity/dm^3		
Breathing rate/breaths min^{-1}		
Ventilation rate/dm^3 min^{-1}		
Oxygen consumption/dm^3 min^{-1}		

> **Examiner tip**
>
> This table uses negative indices for the units: breaths min^{-1} means *breaths per minute*. The oblique line (forward slash or solidus) is used to separate what is being measured from the unit of measurement.

Transport in animals

Module 2

Key words

- open circulatory system
- closed circulatory system
- single circulatory system
- double circulatory system

Hint

A transport system is an example of an organ system – see page 17.

Even small organisms such as *Stentor* (see page 16) do not rely on diffusion alone to transport substances through their cytoplasm. They have vacuoles that move to distribute substances to different areas of their small body. So it is no surprise to find that large organisms cannot rely on diffusion for transport either, as the distances are too great – just think about an elephant or a blue whale. They have very small surface-area-to-volume ratios: many of their cells are far away from exchange surfaces. Animals are very active and require large quantities of oxygen and food – far more than could be delivered by diffusion from those exchange surfaces. As the distances are too great for diffusion alone, multicellular animals have a transport system consisting of a fluid that travels around the whole body, and some sort of pump to move it.

Insects have an *open* circulatory system, as their blood is not enclosed in vessels but circulates in body spaces. Their cells are surrounded by blood. A long, tubular heart circulates the blood through the body spaces. Insect blood has no haemoglobin and no red blood cells. This is because the gas exchange system delivers oxygen and carbon dioxide direct to the tissues in tiny, air-filled tubes. The transport systems in fish and mammals are *closed* circulatory systems, as the blood flows inside vessels such as arteries, veins and capillaries. Their cells are surrounded not by blood, but by **tissue fluid** (see page 27).

✓*Quick check 1*

Fish have a **single circulatory system**. Blood flows through the heart once during one complete circuit of the body. Blood flows from the heart to the gills and then to the rest of the body before returning to the heart.

Head and arms
Lungs
Pulmonary artery
Pulmonary veins
Vena cava
Aorta
Right ventricle
Left ventricle
Stomach and intestines
Liver
Kidneys and legs

Deoxygenated blood
Oxygenated blood

Double circulation in a mammal

Mammals have a **double circulatory system** – blood flows through two circuits:

- **pulmonary** circuit – from the heart to the lungs and back
- **systemic** circuit – from the heart to the rest of the body and back.

When the heart pumps blood, it creates pressure that forces the blood through blood vessels. It is difficult to push blood through the vessels because they exert a resistance to the flow of blood. To overcome this resistance, the heart gives the blood considerable pressure (see page 24).

Blood vessels

Substances such as oxygen, carbon dioxide and glucose are exchanged between blood and tissues through the walls of capillaries. Blood flows from the heart through arteries to reach capillaries and then returns to the heart inside veins. All blood vessels are lined by squamous epithelial cells forming a layer known as endothelium. These endothelial cells help to maintain the uninterrupted flow of blood through the circulation.

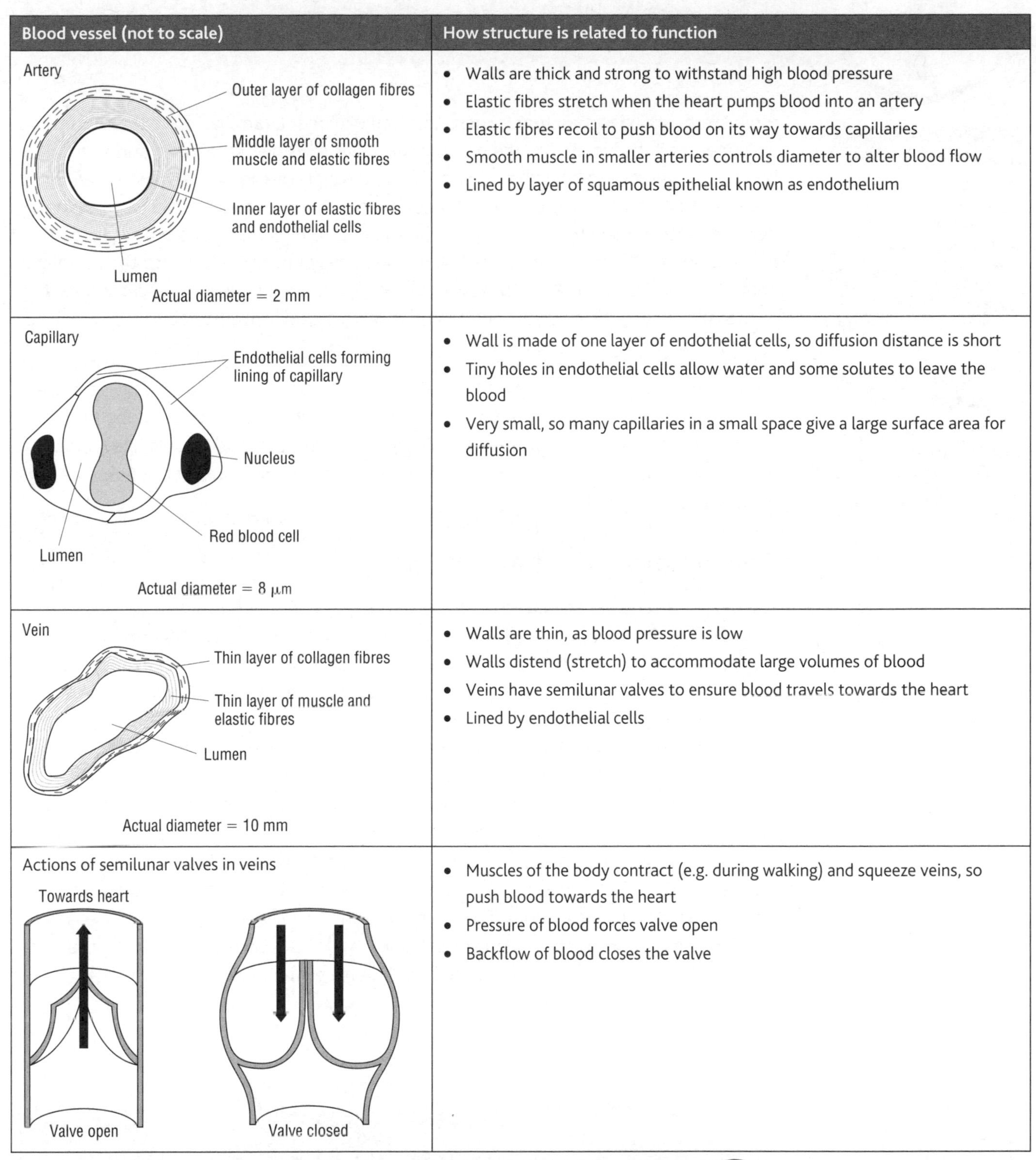

Blood vessel (not to scale)	How structure is related to function
Artery Outer layer of collagen fibres Middle layer of smooth muscle and elastic fibres Inner layer of elastic fibres and endothelial cells Lumen Actual diameter = 2 mm	• Walls are thick and strong to withstand high blood pressure • Elastic fibres stretch when the heart pumps blood into an artery • Elastic fibres recoil to push blood on its way towards capillaries • Smooth muscle in smaller arteries controls diameter to alter blood flow • Lined by layer of squamous epithelial known as endothelium
Capillary Endothelial cells forming lining of capillary Nucleus Red blood cell Lumen Actual diameter = 8 μm	• Wall is made of one layer of endothelial cells, so diffusion distance is short • Tiny holes in endothelial cells allow water and some solutes to leave the blood • Very small, so many capillaries in a small space give a large surface area for diffusion
Vein Thin layer of collagen fibres Thin layer of muscle and elastic fibres Lumen Actual diameter = 10 mm	• Walls are thin, as blood pressure is low • Walls distend (stretch) to accommodate large volumes of blood • Veins have semilunar valves to ensure blood travels towards the heart • Lined by endothelial cells
Actions of semilunar valves in veins Towards heart Valve open Valve closed	• Muscles of the body contract (e.g. during walking) and squeeze veins, so push blood towards the heart • Pressure of blood forces valve open • Backflow of blood closes the valve

QUICK CHECK QUESTIONS

✔Quick check 2, 3, 4

1 Explain why large, multicellular animals need a transport system.

2 Calculate the magnification of the cross-sections of the artery, capillary and vein in the diagrams above.

3 Make a table to compare the structure and functions of the three types of blood vessel. Use these column headings:

Feature	Artery	Capillary	Vein

4 Explain how the structure of arteries, capillaries and veins is related to their functions.

Hint

When you answer quick check question 2, look back to page 2 to see how to do this calculation.

The heart and cardiac cycle

Module 2

Key words

- myogenic
- cardiac cycle
- sinoatrial node
- atrioventricular node
- Purkyne tissue
- ECG

Examiner tip

Examination questions will give blood pressure in kilopascals (kPa).

Put a stethoscope to someone's chest, and you will hear a familiar noise – sometimes described as 'lub-dup'. This is the noise of the valves closing in the heart during a heartbeat. One beat of the heart pumps blood through the pulmonary and systemic circuits. The heart has two pumps working in series. The right side of the heart pumps **deoxygenated** blood to the lungs through the pulmonary artery at a blood pressure of about 24 mmHg (3.2 kPa). The left side pumps **oxygenated** blood into the aorta at about 120 mmHg (15.8 kPa). The flow of blood through the heart is intermittent as it pushes blood into the arteries and then refills with blood from the veins.

The wall of the left ventricle is much thicker than that of the right ventricle. This is because it contracts to force blood into the aorta at high pressure, as the blood in the systemic circuit meets much more resistance to flow. The lungs receive blood from the right ventricle. The lungs are very spongy, and here the blood vessels allow blood to flow easily for maximum exchange of gases in the alveoli. So here the resistance is nowhere near as great.

Coordinating the heart beat

The heart is made of **cardiac muscle**, which is **myogenic** (it contracts of its own accord). The **sinoatrial node** (SAN), which is in the wall of the right **atrium**, is the heart's natural pacemaker. It sends out electrical impulses to start each heartbeat, which is recorded as the P wave in an **electrocardiogram** (see the diagram of an ECG trace on page 25). The impulses spread across the atrial muscle, which contracts first. There is a ring of fibrous tissue between the atria and the ventricles that prevents impulses reaching the ventricles. Impulses reach the **atrioventricular node** (AVN), which is between the atria and ventricles. This is like a relay station that delays impulses for a short while so the ventricles do not contract too soon. The AVN sends impulses along special, fast-conducting muscle cells in the **septum** called **Purkyne tissue**, to the base of the heart. Muscle at the base of the ventricles contracts first to push the blood upwards into the arteries. This electrical activity is recorded as the QRS waves on an ECG. The relaxation of the ventricles follows, and this is recorded as the T wave. The SAN determines the heart's natural rhythm. Nerves to the heart change the heart rate according to circumstances.

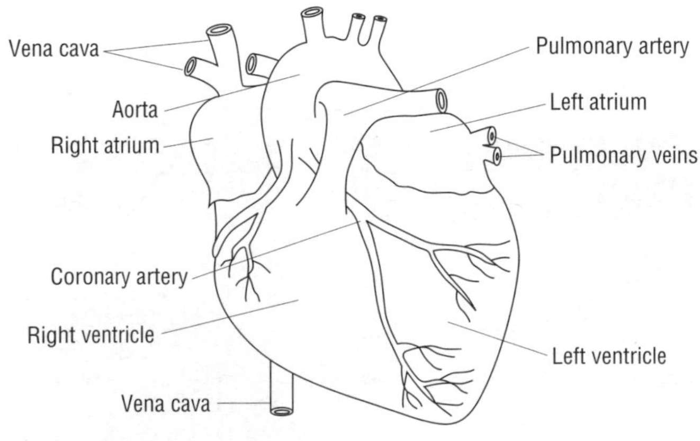

External view of the human heart showing the four chambers, main blood vessels, and coronary arteries that supply blood to heart muscle

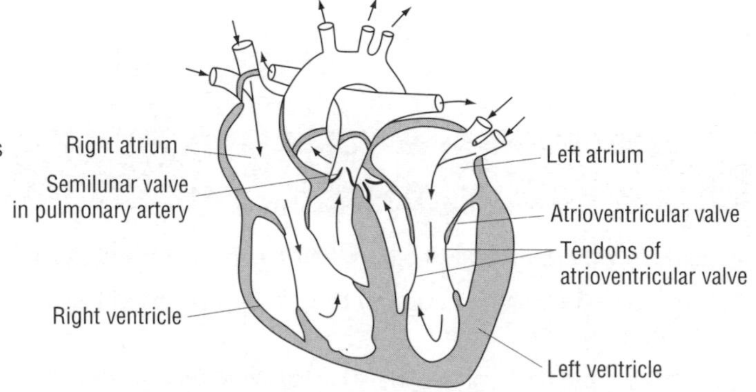

Vertical section through the heart to show the internal structure. The left and right atria have much thinner walls than the ventricles because the atria contract to push blood into the ventricles at low pressure, to ensure that they are full

The cardiac cycle

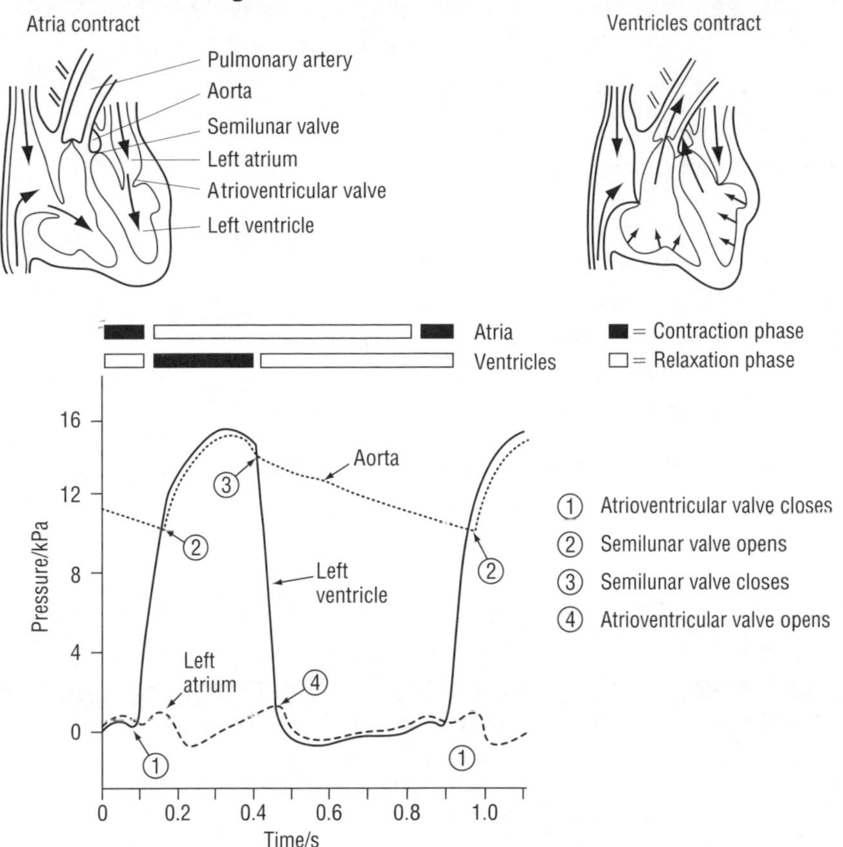

The cardiac cycle, showing changes in blood pressure in the left atrium, left ventricle and aorta during one heartbeat and the beginning of the next

Hint

Move a ruler from left to right across the graph and follow the changes that occur.

Module 2

Quick check 1, 2, 3

Monitoring the heart

Electrocardiograms record electrical activity of the heart. These are taken if doctors suspect there is a problem with the way the heart functions. The figure shows a normal ECG trace. An irregular pattern of waves suggests that there may be a problem with the SAN.

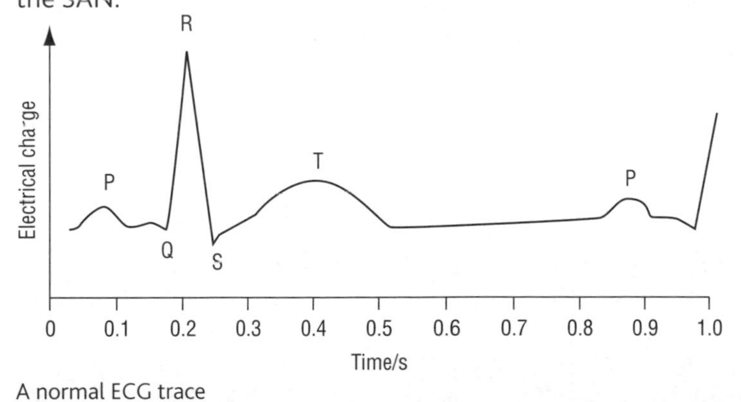

A normal ECG trace

Quick check 4

QUICK CHECK QUESTIONS

1 State the length of time taken by one heartbeat in the graph of the cardiac cycle. Use your answer to calculate the heart rate in beats per minute.

2 Explain why the atrioventricular valve closes at 1 and opens at 4.

3 Explain why the semilunar valve opens at 2 and closes at 3.

4 Explain what information can be gained from an ECG.

Hint

Study the graph of the cardiac cycle carefully before answering quick check questions 1–3.

Blood, tissue fluid and lymph

Key words

- plasma
- tissue fluid
- lymph
- pressure filtration

Blood is a suspension of red and white cells and platelets in plasma. When left to settle, or spun in a centrifuge, blood separates into these three components.

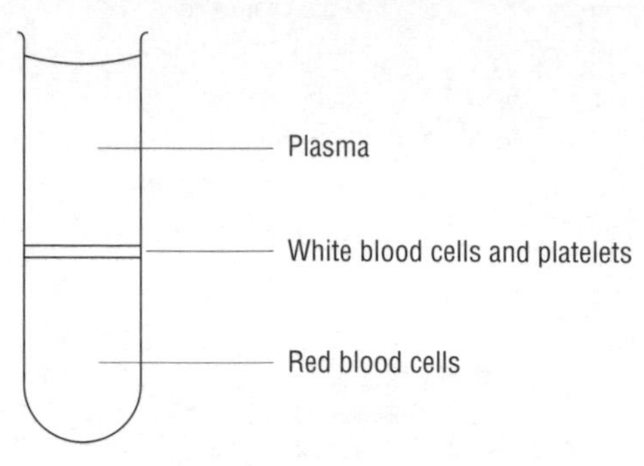

The composition of blood

Tissue fluid is a colourless fluid that is formed from blood plasma by pressure filtration through capillary walls. It surrounds all the cells of the body, and all exchanges between blood and cells occur through it.

Lymph is tissue fluid that has drained into **lymphatic** vessels. It passes through lymph nodes, where it gains white cells and antibodies. Lymphatic vessels absorb hormones from some endocrine glands and fat in the small intestine.

The table below shows the structure of the blood cells as seen through the light microscope.

Cell	Relationship between structure and function
Red blood cell (7 μm)	• Biconcave disc shape gives a large surface area for diffusion of oxygen and carbon dioxide
	• No organelles, so cytoplasm is full of haemoglobin
	• Elastic membrane allows cells to change shape as they squeeze through capillaries and restore shape when they enter veins
Phagocyte (neutrophil) (9 μm)	• Phagocytosis – bacteria engulfed in vacuoles and digested
	• Large number of lysosomes for digestion of bacteria
	• Lobed nucleus to help squeeze through gaps between cells in capillary walls
Lymphocyte (5.5 μm)	• Some lymphocytes develop into plasma cells that have a large quantity of rough endoplasmic reticulum for fast production of antibodies (see page 74)

✓ *Quick check 1, 2, 3*

This table summarises the differences between blood, tissue fluid and lymph.

Component	Blood	Tissue fluid	Lymph
Red blood cells	✓	✗	✗
White blood cells	✓	Some	✓
Water	✓	✓	✓
Plasma proteins	✓	Very few	Very few
Sodium ions	✓	✓	✓
Glucose	✓	✓	Very little
Antibodies	✓	✓	✓
Fats	✓	Some	✓ especially after a meal

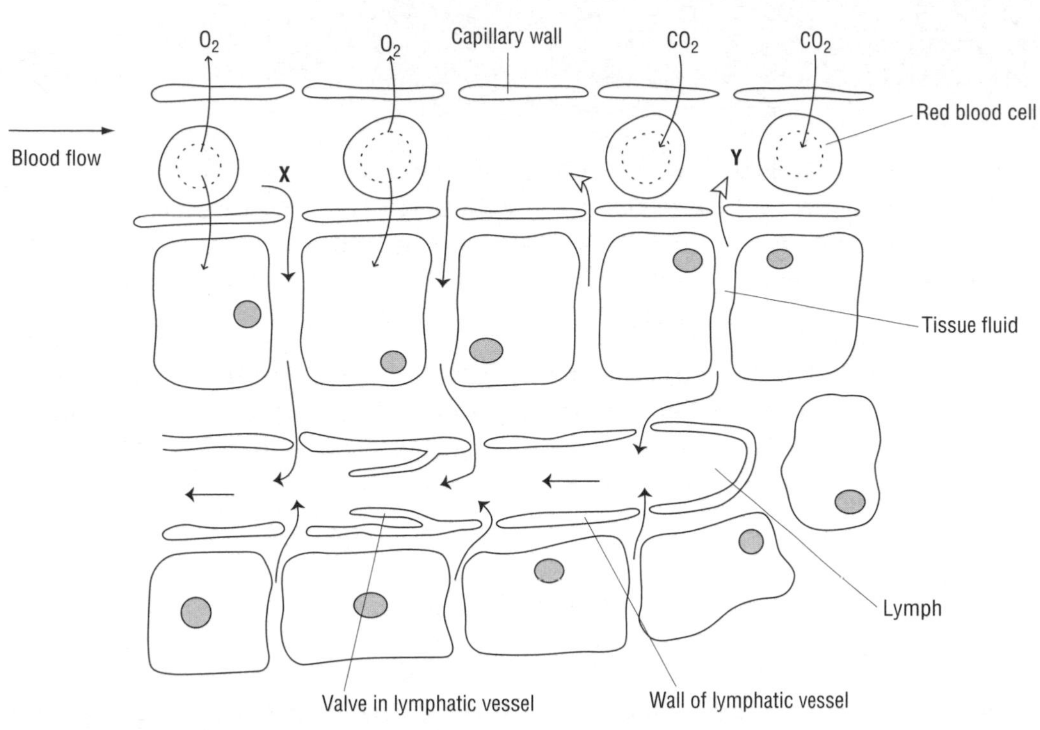

Tissue fluid bathes all the cells of the body. It is formed from blood plasma and drains into lymphatic vessels to form lymph

Examiner tip

Remember that substances are exchanged between tissue fluid and cells by diffusion, facilitated diffusion, active transport and osmosis. See pages 8 to 9 to remind yourself of these processes.

Module 2

Blood enters capillaries at a relatively high pressure. This forces small molecules, such as water, ions and glucose, through the small holes in the walls of the endothelial cells to form tissue fluid. This is called pressure filtration. Much of the water – but not all – returns to the blood by osmosis further along the capillaries, where the pressure has dropped. Excess fluid returns to the blood through the lymphatic vessels, ensuring that tissues do not fill with too much fluid. When they do, a person suffers from oedema, which can be very dangerous if it happens in organs such as the lungs. Lymphatic vessels eventually join together to form two large ducts that empty into the blood near the heart.

Hint

Pressure filtration is important in the kidneys, where it helps to form urine. You will learn more about this in your A2 course.

Red and white blood cells are produced by stem cells in bone marrow. Stem cells divide by mitosis to form cells that differentiate into the cell types in the table opposite. Most of the white blood cells referred to in the table are lymphocytes, which enter the lymph as it flows through lymph nodes (see page 64). They then circulate round the body in the blood. Lymphocytes secrete antibodies, which explains why they are in all three fluids.

✓Quick check 4–8

QUICK CHECK QUESTIONS

1 Calculate the magnification of the red blood cell shown on page 26.
2 Explain how the structure of red blood cells is related to their function.
3 Draw a table to compare red blood cells, neutrophils and lymphocytes.
4 Explain why red blood cells are not found in tissue fluid, but white cells are.
5 Explain why the walls of capillaries are composed of squamous epithelial cells.
6 State how the following are exchanged between tissue fluid and cells: oxygen, water, glucose and ions.
7 What causes blood to have a pressure?
8 Explain why water leaves a capillary at one end (X on the diagram above) and returns at the other end (Y).

Hint

Before answering quick check question 6, look at pages 8 and 9.

Hint

Before answering quick check question 8, remind yourself about water potential gradients.

Haemoglobin and gas transport

Key words

- haemoglobin
- partial pressure
- Bohr shift
- affinity

Hint

Haemoglobin is a **globular protein** made of four **polypeptides** (2α and 2β). Each of these has a **haem** group for binding oxygen – see page 42 for details of the structure of haemoglobin.

Examiner tip

Do not confuse carbaminohaemoglobin with *carboxyhaemoglobin*, which forms when carbon *monoxide* (rather than carbon *dioxide*) combines with haemoglobin (see page 78).

✓*Quick check 1*

Oxygen is not very soluble in water. If we did not have haemoglobin in red blood cells to transport oxygen, the blood would carry about 0.3 cm^3 of oxygen in every 100 cm^3 (the volume of oxygen that can dissolve in water). When blood leaves the lungs, it carries far more than this – every 100 cm^3 of blood carries 20 cm^3 of oxygen. Almost all of this is combined with haemoglobin.

There are over 280 million molecules of haemoglobin packed into each red blood cell, and each haemoglobin molecule can carry up to four molecules of oxygen. When blood flows through the capillaries in the lungs, haemoglobin forms **oxyhaemoglobin**. A molecule of oxygen combines with each haem group:

$$\mathbf{Hb + 4O_2 \rightarrow HbO_8}$$

In the tissues, this reverses and oxyhaemoglobin dissociates to give up oxygen:

$$\mathbf{HbO_8 \rightarrow Hb + 4O_2}$$

Haemoglobin also transports carbon dioxide. As blood flows through tissues, some carbon dioxide molecules react with free amino ($-NH_2$) groups at the ends of the α (alpha) and β (beta) polypeptides to form a compound known as **carbaminohaemoglobin**.

Dissociation curves

Dissociation curves show how efficient haemoglobin is at absorbing oxygen in the lungs and delivering oxygen to tissues. Samples of blood are exposed to different mixtures of oxygen and nitrogen, and shaken to ensure that haemoglobin absorbs as much oxygen as possible. **Partial pressure** (measured in kilopascals, kPa) is the pressure exerted by oxygen in this mixture. The percentage saturation is calculated as the percentage of the *maximum* quantity of oxygen that haemoglobin absorbs.

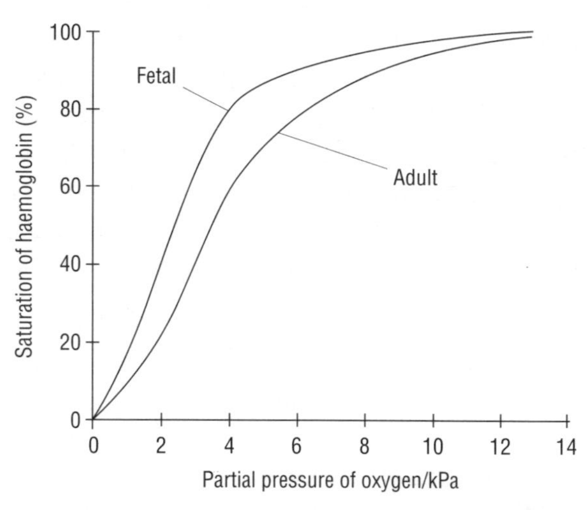

Dissociation curves for adult and fetal haemoglobin

	Partial pressure of oxygen/kPa	Saturation of haemoglobin with oxygen/%
Lungs (at sea level)	13.0	98
Respiring tissues	5.0	70
Actively respiring tissues (e.g. during exercise)	3.5	50
Maternal blood in the placenta	4.0	60
Fetal blood in the placenta	4.0	80

Put a ruler along the vertical axis on the graph above. Move the ruler along the graph from 0 to 14 kPa. An important feature is the sigmoid or S-shape of the curve. Position the ruler on the graph to confirm the readings in the table. Start with the ruler on 14 kPa. Confirm the observations that follow.

Hint

You can use the dissociation curves to predict the saturation of the blood with oxygen in different parts of the circulation.

- Haemoglobin is fully saturated at the partial pressure of oxygen in the lungs. It becomes fully loaded with oxygen as it passes the gas exchange surface (the alveoli). When fully saturated, almost all the haemoglobin is in the form of *oxyhaemoglobin*.
- As it flows through the tissues, oxyhaemoglobin responds to low partial pressures of oxygen by dissociating (giving up some of its oxygen). Oxygen diffuses through capillary walls and tissue fluid to respiring cells.
- As tissues use up oxygen during exercise, oxyhaemoglobin dissociates even more.
- The S-shape of the curve shows that oxyhaemoglobin responds to small decreases in oxygen concentration in the tissues by giving up a lot of oxygen.
- Fetal blood has a higher **affinity** for oxygen than adult blood. This means that oxygen diffuses from maternal blood to fetal blood across the placenta, even though the partial pressures are similar.

Quick check 2, 3, 4

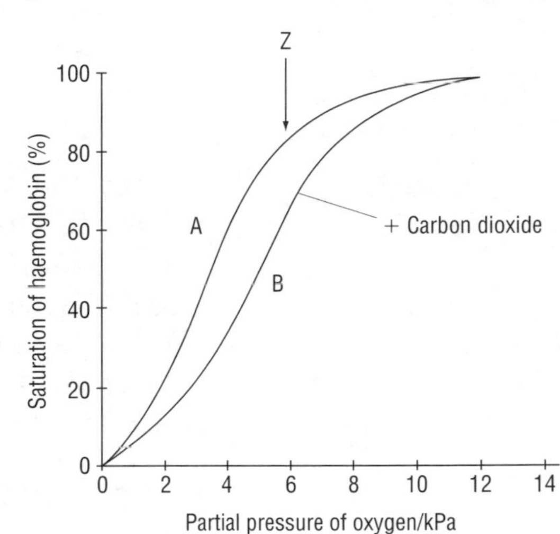

When cells respire, they produce carbon dioxide. The graph on this page shows what happens when carbon dioxide is added to the gas mixture. Take a ruler and put it at the point marked Z on the graph. When there is more carbon dioxide in the mixture, haemoglobin is less saturated with oxygen.

Now put the ruler parallel with the horizontal axis and see that curve B is to the right of A. This effect of carbon dioxide on the dissociation curve is the **Bohr shift** or Bohr effect.

Carbon dioxide interacts with haemoglobin to cause it to give up its oxygen. This is good news, because it means that haemoglobin delivers more oxygen to those tissues that are respiring quickly, such as muscles during exercise. This happens because an enzyme in red blood cells, carbonic anhydrase, catalyses the reaction between water and carbon dioxide:

$$H_2O + CO_2 \xrightleftharpoons{\text{carbonic anhydrase}} H_2CO_3 \rightleftharpoons H^+ + HCO_3^-$$

Haemoglobin absorbs the hydrogen ions that form inside the red blood cells, which causes them to lose the oxygen molecules they are carrying.

Quick check 5

QUICK CHECK QUESTIONS

1. Describe how haemoglobin transports oxygen and carbon dioxide.
2. The graphs in this spread show the results of experiments in which gas mixtures have been bubbled through samples of blood. State *two* factors that should be kept constant when carrying out these experiments.
3. Explain the advantage of the S-shaped dissociation curve for haemoglobin.
4. Use the first graph (page 28) to explain what is meant by the following phrase: 'fetal haemoglobin has a higher affinity for oxygen than does adult haemoglobin'.
5. Describe and explain the Bohr effect.

Transport in plants

Key words

- assimilates
- mass flow
- apoplast
- symplast
- plasmodesmata
- endodermis

Hint

See page 17 to check where these tissues are located inside plant organs.

Transport in plants

Plants use photosynthesis to convert light energy to chemical energy in compounds such as carbohydrates, fats and proteins. The raw materials in this process of assimilation are simple inorganic substances, such as carbon dioxide, water and ions.

Transport tissues

Plants are large organisms, and sites of production of carbohydrates (leaves) are often a long way from sites of use (e.g. roots). Diffusion would be inadequate for such long distances, so plants have transport tissues:

- in **xylem tissue** water and ions move from roots to stems, leaves, flowers and fruits
- in **phloem tissue** sucrose and other **assimilates** travel upwards and downwards.

Movement in xylem and in phloem is by **mass flow**. Everything travels in the same direction within each column of xylem or phloem cells.

Plants do not have a transport tissue for oxygen and carbon dioxide. As plants have a much lower metabolic rate and a larger surface-area-to-volume ratio than animals, diffusion through air spaces to and from cells is sufficient. This is possible because oxygen, for example, diffuses 10 000 times faster through air than through water or cells. There are air spaces between cells in all plant organs, so gases can reach the centre of even a thick root. The cell surfaces exposed to air spaces form a large gas exchange surface. You can see this in the figure below.

The pathway of water through a plant

Water is absorbed into root hair cells by osmosis. Root hair cells are well adapted for absorption as they have:

- a large surface area
- thin cell walls
- lower water potential than soil water
- carrier molecules in cell membranes for absorption of ions.

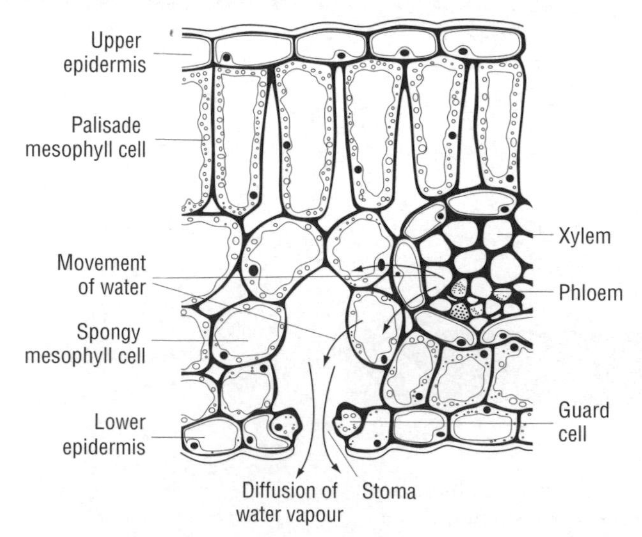

Cross-section of part of a leaf blade (compare this with the leaf section on page 17, which is a plan drawing showing tissues, not cells)

Most of the water absorbed takes the pathway of least resistance *along cell walls* in the **cortex** of the root until it reaches the **endodermis**. Here, it cannot continue because of the **Casparian strip** – a region of impermeable material (suberin) within the cell walls. Water has to move *through the endodermal cells* to enter the xylem. Water moves across the endodermis down a water potential gradient. To achieve this gradient, cells of the endodermis use active transport to pump ions from the cortex into the xylem, giving a *lower* water potential in the xylem than in the cortex.

- The **apoplast pathway** is the pathway *along cell walls* and in the spaces between cells. Water passes along the apoplast pathway when crossing the cortex of the root (see the diagram on page 31).
- The **symplast pathway** is the pathway *through cells*. Water flows from cell to cell through **plasmodesmata**, microscopic cytoplasmic connections between cells, without crossing cell walls.

Quick check 1

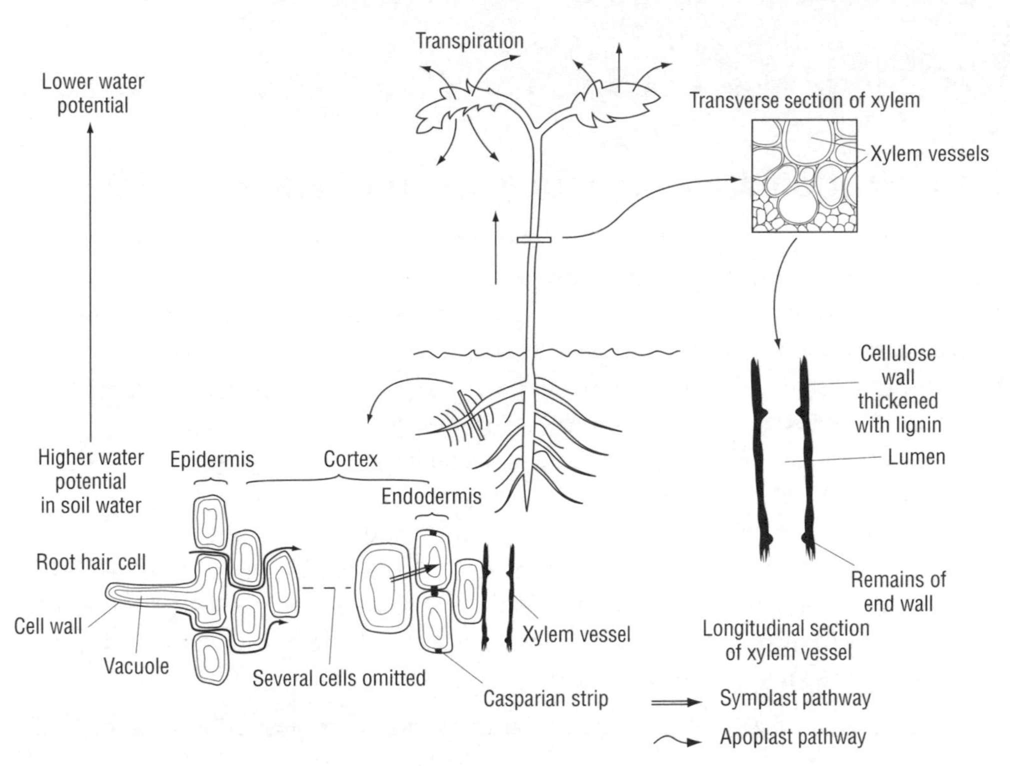

The pathway taken by water from the soil to the atmosphere

Quick check 2

Water travels up the xylem from roots into leaves. In leaves, water passes from the xylem into the cell walls and cytoplasm of mesophyll cells. Much of the water passes *along the cell walls* and then evaporates to form water vapour (see page 32). Some water passes from cell to cell through **plasmodesmata**. Most water vapour passes out of leaves through the **stomata**, but some passes through the cuticle.

Examiner tip

Remember – water travels *down* its water potential gradient.

Plants lose huge amounts of water vapour to the atmosphere by **transpiration**:

- water evaporates from moist cell walls
- water vapour diffuses from air spaces inside leaves to the atmosphere.

Palisade and spongy mesophyll cells have a very large internal surface. The carbon dioxide concentration in air is low (0.04% or 400 parts per million), so plants need large gas exchange surfaces to absorb enough carbon dioxide for photosynthesis.

The air inside leaves is always saturated with water vapour. Usually the air outside is less saturated, so a concentration gradient for water vapour exists between the air spaces and the outside. Water vapour diffuses down this humidity gradient. The pathway with the least resistance is through the stomata. Stomata are open during the day to allow carbon dioxide to diffuse in. At night, guard cells close the stomata, so the rate of water loss decreases. Environmental factors, such as wind speed, temperature, light intensity and humidity, influence transpiration rate.

Quick check 3

QUICK CHECK QUESTIONS

1 Explain how roots absorb water from the soil.
2 Describe the pathway taken by water as it moves through a plant from the soil to the atmosphere.
3 Explain the function of the Casparian strip in the endodermis of the root.

Transport in the xylem: transpiration

Key words

- transpiration
- cohesion
- adhesion
- cohesion–tension
- lignin
- potometer
- xerophyte

Hint

Note the importance of hydrogen bonds – these are described on page 38.

Examiner tip

When *explaining* the movement of water in plants, start with water loss from leaves, as here.

Hint

The development of xylem vessel elements is an example of differentiation.

Quick check 1

Transpiration drives the movement of water in plants

The loss of water from leaves by transpiration causes water to travel upwards through the plant by mass flow. The mechanism, called **cohesion–tension**, works as follows.

- Water molecules 'stick' to each other by **hydrogen bonds**. This is called **cohesion**.
- Water evaporates from mesophyll cells, and this lowers their water potential.
- Water moves along gradients of water potential from the high water potential in the xylem to the lower water potential in the mesophyll cells.
- Continuous columns of water molecules are pulled from the xylem towards the mesophyll as a result of forces of cohesion.
- The pull on these columns extends all the way to the roots and into the soil.
- Water molecules also form hydrogen bonds to **cellulose** in cell walls (**adhesion**). This helps maintain columns of water in the xylem when there is little transpiration.
- The energy for water movement comes from thermal energy evaporating water from the mesophyll surface.

This mass flow of water through the plant is known as the transpiration stream.

Xylem vessels are adapted for transport of water

Meristematic cells in the lateral **cambium** divide by mitosis to form cells called **xylem vessel elements**. Enzymes in the cytoplasm of these long cells make cellulose and **lignin**, which are deposited in the cell wall by exocytosis. The end walls are not thickened this way; they remain very thin and, as water starts to leave the vessel elements, they break and are carried upwards in the transpiration stream along with cell contents, leaving an empty column – a xylem vessel (see the diagram on page 31).

Lignin prevents xylem vessels collapsing inwards under the tension that develops when rates of transpiration are high. It also waterproofs the vessels to keep water inside. Parts of the walls, known as pits, are not lignified so that water can move out into the surrounding tissues.

Measuring transpiration

We can find out how much water plants take up through their roots and lose into the atmosphere as water vapour by carrying out a variety of investigations.

The diagram on the left shows two ways to measure water uptake: you can follow the fall in water level or the loss in mass. The disadvantage of these methods is that it takes a long time to obtain results.

You can make a better potometer from simple apparatus (see figure on next page) and use it to measure the uptake of water by leafy shoots.

1. Cut a shoot under water and place it in rubber tubing, still under water. This prevents air getting into the stem and blocking the xylem vessels so that water cannot flow.
2. Allow time for the plant to adjust to the surroundings, and keep environmental conditions constant (light intensity, humidity, temperature, air movement).
3. Time how long it takes the meniscus to move a set distance.
4. Reset the meniscus with water from the syringe, and take more readings until the rate of uptake is fairly constant.
5. Continue to record results, then calculate the mean rate of uptake. Exclude anomalous results from your calculation.

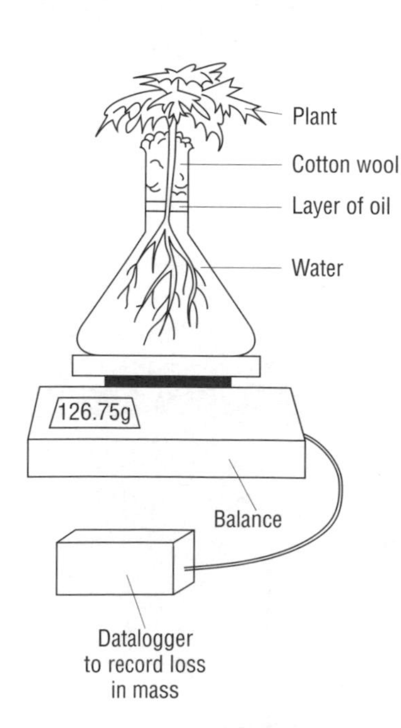

Quick check 2

As most water absorbed by the plant is lost to the atmosphere, we have assumed that the rate of uptake is the rate of transpiration. However, the plant uses water as a raw material in photosynthesis and for keeping cells turgid. A potometer on a sensitive balance can record very small changes in mass, that is, the rate of transpiration, at the same time as the rate of water uptake (see the graph below).

- rates of water uptake and water loss are highest around midday
- rates of transpiration are low at night because stomata are closed
- water is absorbed at night because so much water has been lost during the day that more is needed to maintain the turgidity of cells.

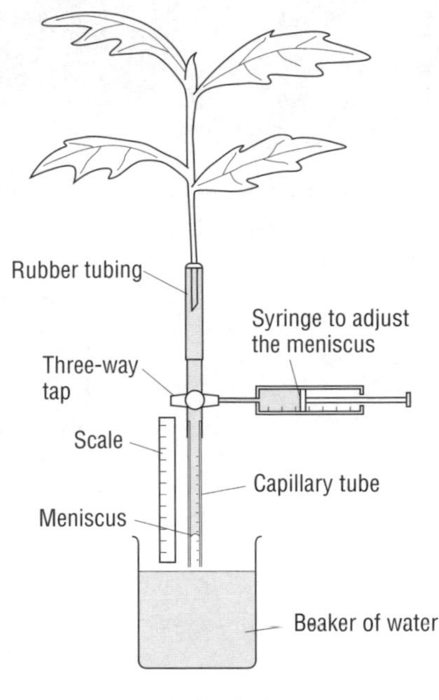

A simple potometer

Environmental factors influence transpiration rate

Potometer experiments show that four factors influence the rate of transpiration.

Stomata are usually closed in the dark. As **light intensity** increases, they open, so more water vapour can escape. Above a certain light intensity the stomata cannot open any wider, so the rate of transpiration remains constant.

The **humidity** of the air around a plant determines the concentration gradient between the air spaces in the leaf and outside. If the air outside is dry, the gradient is steep. If the air is saturated (like the air in the leaf), there is no gradient and no net loss of water vapour through the stomata.

Temperature determines rates of evaporation inside leaves and also the water-holding capacity of the air outside the leaf. High rates of transpiration occur on hot days.

When there is no **air movement** around a leaf, water vapour molecules collect around the leaf surface so the air is saturated. Little or no transpiration occurs. Air blowing over the surface of a leaf carries water vapour away, so rates of transpiration are high.

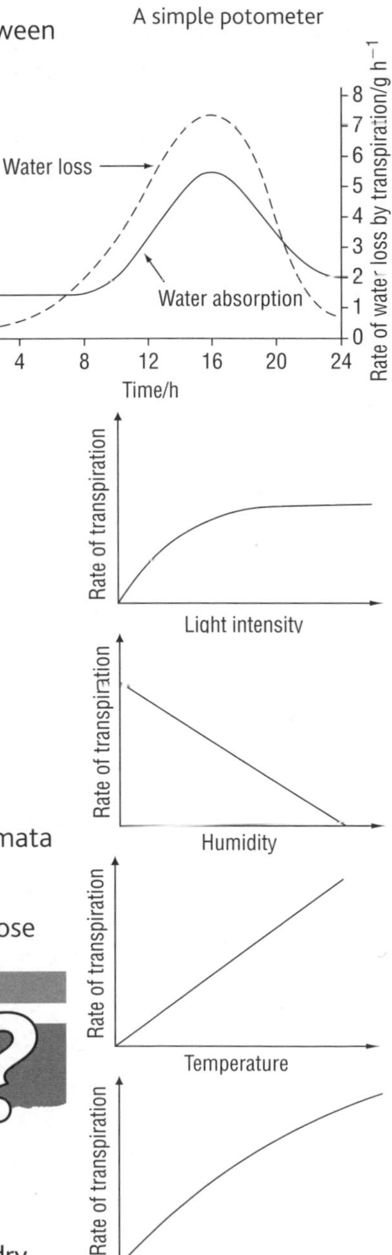

Xerophytes

These plants can survive in dry places, such as deserts and sand dunes. They have adaptations to reduce transpiration:

- small leaves to reduce the surface area
- thick leaves to reduce the surface-area-to-volume ratio
- stomata set deep in the leaf, so they are in a depression full of water vapour
- thick, waxy cuticles to reduce water loss through the epidermis.

Marram grass, *Ammophila arenaria*, grows on sand dunes. As well as having the adaptations above, its leaves roll into cylindrical shapes in dry weather. All the stomata face inwards, so they are surrounded by air saturated with water vapour.

The stomata of some desert plants open at night to collect carbon dioxide, then close during the day. This reduces water loss considerably.

Quick check 3

QUICK CHECK QUESTIONS

1. Explain how xylem vessels are adapted for transport of water.
2. Describe the steps you should take when using a potometer to measure rates of transpiration at different air speeds. Explain any precautions that are necessary.
3. Explain how the leaves of marram grass are adapted to help the plant survive in hot, dry conditions.

Transport in the phloem

Key words

- translocation
- sources
- sinks
- sieve tubes
- companion cells

Most photosynthesis in plants occurs in leaves, which use carbon dioxide and water as raw materials for making three-carbon sugars known as trioses. These substances are used to make many different compounds. All the reactions of photosynthesis occur in chloroplasts. These organelles convert simple sugars into amino acids, using nitrate ions (NO_3^-) or ammonium ions (NH_4^+), and make chlorophyll using magnesium ions (Mg^{2+}). In the cytoplasm of leaf cells, trioses are converted into glucose and fructose, and these are combined to make sucrose. Sucrose is the main carbohydrate used by plants for transport because it is unreactive – far less reactive than glucose or fructose, which are reducing sugars (see page 44). Compounds that plants have made from simple raw materials are called **assimilates** (they are produced by **assimilation**). Many of these assimilates are exported from leaves to the rest of the plant in the **phloem**.

Sources and sinks

The transport of assimilates in phloem tissue is called **translocation**, which means 'from one place to another'. Leaves are known as **sources** as they load assimilates into the phloem. Assimilates are transported to other parts of the plant, which use them to provide energy or materials for synthesising macromolecules such as cellulose and proteins. These locations in the plant, such as meristems, roots, stems, flowers, fruits and seeds, are referred to as **sinks**.

Movement in the phloem is an active process

Sucrose and other assimilates travel throughout a plant in phloem sieve tubes, which are made from cells called **sieve tube elements**. Alongside sieve tubes are **companion cells**. Mesophyll cells in the leaf are close to veins containing sieve tubes. Sucrose and other assimilates are loaded into sieve elements by companion cells.

Sucrose travels to the phloem companion cells in two ways:

- from cell to cell through narrow tubes of cytoplasm that penetrate cell walls (known as plasmodesmata)
- along cell walls in the mesophyll.

Carrier proteins in the cell surface membranes of companion cells actively pump sucrose into the cytoplasm. Mitochondria in the companion cells provide ATP for this pumping. From here, sucrose passes through plasmodesmata into sieve tube elements.

The accumulation of sucrose and other solutes, such as amino acids, in sieve tube elements lowers the water potential so that water diffuses in by osmosis from adjacent cells and from the xylem. This creates pressure in the sieve tube elements, causing the liquid (phloem sap) to flow along the sieve tubes out of the leaf.

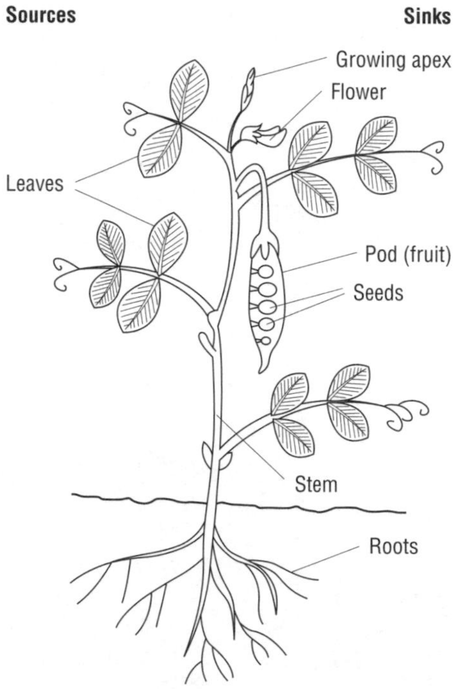

Sources
Sinks
Growing apex
Flower
Leaves
Pod (fruit)
Seeds
Stem
Roots

Sources and sinks in a pea plant

Phloem sieve tube elements are adapted for transport as they have:

- end walls with sieve pores, allowing phloem sap to flow freely
- little cytoplasm (which would impede the flow of sap)
- plasmodesmata, to allow assimilates to flow in from companion cells.

Sieve tube elements differ from xylem vessel elements because they are alive: they have some cytoplasm with organelles although they do not have nuclei. They are surrounded by a cell surface membrane to prevent loss of sucrose. They are not lignified, as they do not need to withstand the same forces of tension that exist in xylem. The function of sieve plates may be to prevent sieve tubes from bursting. ✓Quick check 1

Sucrose is *unloaded* at sinks. It is not known how this is done, but it seems likely that an enzyme breaks down sucrose to glucose and fructose, which are taken up by cells and respired, or stored as starch, which is insoluble. This reduces the solute concentration of phloem sap, so water diffuses out of the phloem, thus decreasing the pressure. This maintains a pressure gradient from source to sink so that sap keeps flowing in the phloem. ✓Quick check 2, 3

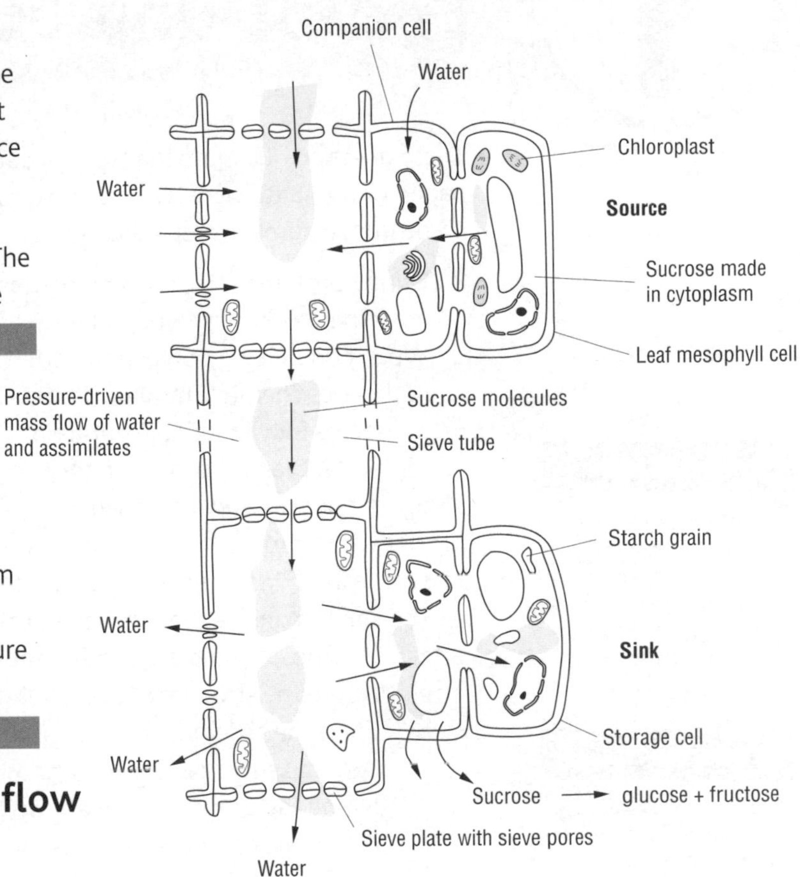

Evidence for and against mass flow

Evidence for

- Companion cells have many mitochondria to produce ATP for loading of sucrose.
- The pH of companion cells is *higher* than that of other cells because they pump protons (H^+) out of their cells as part of the loading process.
- Phloem sap moves much more quickly than could occur by diffusion – up to 10 000 times faster.
- When aphids (small, sap-sucking insects) insert their stylets (mouthparts) into sieve tubes, phloem sap flows into the aphid under pressure.
- Pressures recorded in phloem are more than adequate to move phloem sap within sieve tubes.

Evidence against

- Sieve tubes contain a special protein – called P protein. This seems to have no function in mass flow in the phloem, although it may act to block sieve pores when sieve tubes are damaged and prevent loss of valuable sucrose.
- Some plants have thin protein strands extending through the sieve pores – these have no function in mass flow.
- Phosphates and sucrose can move at different rates in the same vascular bundle in the phloem.

✓Quick check 4

QUICK CHECK QUESTIONS

1 Make a table to compare the structure and functions of xylem vessels and phloem sieve tubes.

2 Describe how phloem sieve tube elements are adapted for the transport of sucrose.

3 Explain how sucrose is loaded into a sieve element at a source and how it may be unloaded at a sink.

4 Explain how phloem sap is moved from a source to a sink by mass flow.

Hint

Before answering quick check question 1, look back to page 31 for details of xylem structure. Remember to use three columns in your table.

End-of-unit questions

These questions illustrate the types of question that are likely to be on the examination paper for this unit. There is also some advice on examination technique. The answers may come from a number of spreads in this unit. You will find the answers on page 97.

1 During protein synthesis, amino acids are joined to form polypeptides. When complete, a polypeptide often undergoes chemical modification to produce a fully functioning protein molecule. Diagrams (A–D) in the figure opposite show organelles into which protein moves as it is synthesised, modified and secreted from the cell.

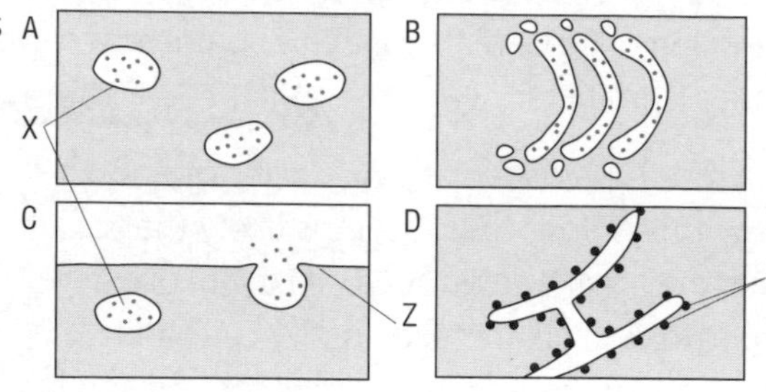

The solution inside the organelles, containing newly synthesised protein, is indicated as dots. Phospholipid bilayers are shown as single black lines.

a Name the structures labelled **X**, **Y** and **Z**. (3)

b State the order in which diagrams A to D should be arranged to show the correct sequence for the synthesis and chemical processing of proteins. (1)

c Name the process that is shown happening in diagram C. (1)

d Name the organelle which is shown in diagram B. (1)

Examiner tip

When you see a new or unfamiliar diagram, concentrate first on the features you recognise.

Examiner tip

Read all parts of the question before you start writing down any answers.

2 Study the following passage carefully and answer the questions below.

It has been possible to develop artificial urinary bladders by growing the different types of differentiated cells that make up a bladder in tissue culture. When large populations of the different cell types have been produced, they are placed on a framework of collagen fibres and develop into a functioning bladder, which can be connected in place of a damaged or defective bladder.

An artificial heart cannot be produced in this way because heart muscle cells rarely, if ever, divide, either in the heart itself or in tissue culture. Much research is taking place so that stem cells can be successfully transferred from culture into a damaged heart. In animal models, the stem cells have been shown to divide and differentiate into new heart muscle cells and new blood vessels. Stem cells for this procedure have been obtained from bone marrow or from an embryo.

a How does the passage show that the urinary bladder is an organ? (1)

b Explain the difference between a stem cell and a differentiated cell. (5)

c Name two types of cell that are formed from stem cells in bone marrow. (2)

d Suggest some problems that must be overcome for stem cell therapy to become an established medical procedure for treating humans. (3)

e State the name given to cells in a plant body that have a function similar to stem cells in a mammalian body. (1)

Examiner tip

When answering questions on a passage, look for ideas or words in the passage at the same time as using your own knowledge.

Examiner tip

Question 2d involves complex issues. Try and write a paragraph that outlines three different problems, for three marks. You can list the problems if you prefer.

3 The diagram at the top of page 37 shows part of a human ribcage, viewed from the side. The layers of muscle fibres arranged between the ribs are labelled **A** and **B.**

a Name the muscle layers **A** and **B**. (2)

b What force causes the alveoli to inflate during inspiration? (1)

c When a person is at rest and the tidal volume and breathing rate are low, what causes the alveoli to partially deflate during exhalation? (1)

When someone is lying on their back, the abdomen rises and falls as they breathe.

d Explain why the abdomen rises and falls. (2)

The ribs of adult humans are attached to the spinal column at an angle, as shown in the diagram opposite.

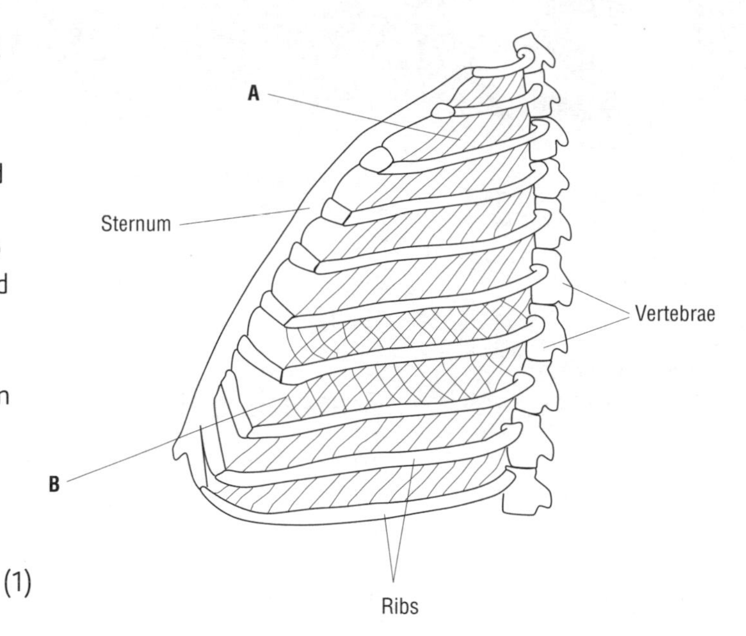

4 A spirometer is a box that is balanced or suspended over water. A known volume of air or oxygen is enclosed inside this box. The box is free to move up and down in the water as a person breathes out and in. The person breathes the trapped air or oxygen by means of a mouthpiece and a system of tubes and valves. A movement sensor records the position of the box. Soda lime granules are used to absorb carbon dioxide from the exhaled breath before it returns to the box.

a Explain why a person must wear a nose clip while using a spirometer. (1)

b State *three* safety precautions that should be taken when using a spirometer. (3)

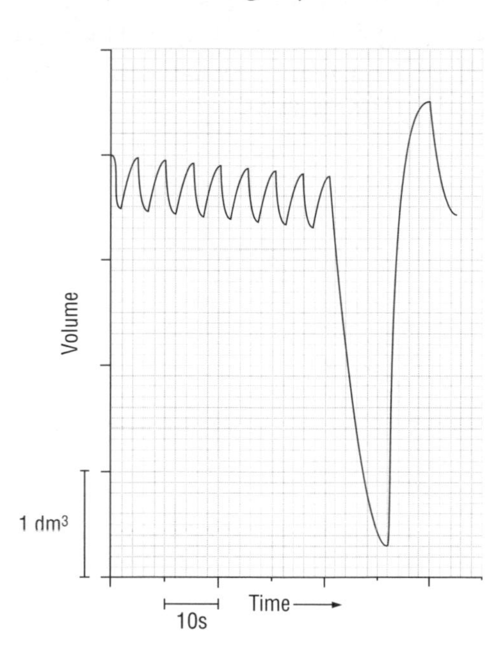

The figure above shows a record of breathing made using a spirometer filled with medical-grade oxygen. Use this recording to answer the following questions.

c State the tidal volume and the vital capacity of the person. (2)

d Explain why the spirometer trace falls steadily during the first 30 seconds. (3)

5 **a** Name *three* organelles found within the cytoplasm of eukaryotic cells. (3)

The solution of proteins and other substances that surrounds organelles is often called the cytosol. This solution is criss-crossed by extremely fine filaments made up of protein subunits. The 3D network of these filaments is called the cytoskeleton. Two types of filament in the cytoskeleton are microfilaments and microtubules.

b **i** State *three* functions of the cytoskeleton in animal cells. (3)

ii Suggest why nerve cells (neurones) have an unusually well-developed cytoskeleton. (1)

iii Suggest why most plant cells have a poorly developed cytoskeleton. (1)

Hint

Remember – the lid of the box goes up when the person breathes out.

Examiner tip

Lay the edge of a ruler along the peaks representing the individual expirations when calculating oxygen consumption.

Examiner tip

Questions 2d, 5bii and 5biii asks you to *suggest*. This means you must work out a reasonable answer using your knowledge – you are unlikely to have learnt the answer to a 'suggest' question.

Water and macromolecules

Key words

- solvent
- monomer
- polymer
- macromolecule
- condensation
- hydrolysis

Life originated in water. The body of a mammal is about 60–70% water. Plant tissues are about 90% water. Many animals and plants live in water. One of water's important features is that it is a good **solvent**. Cytoplasm and body fluids, such as blood plasma, tissue fluid and lymph are aqueous (watery) solutions containing ions and dissolved substances.

Water is a liquid

From its formula (H_2O), water should be a gas at the temperatures we experience on Earth. A heavier molecule with a similar formula, hydrogen sulfide (H_2S), is a gas.

Water is a liquid because of hydrogen bonds between water molecules. The bonds between oxygen and hydrogen are covalent, with electrons shared between them. However, the oxygen atom exerts an attraction for the electrons in the covalent bonds, making the oxygen slightly negatively charged (δ–) and the hydrogen slightly positively charged (δ+). The symbol δ (delta) indicates a slight charge. The attraction between δ+ and δ– is a hydrogen bond, and each water molecule may form up to four of these to make a cluster. In water, the clusters break and re-form all the time.

✔Quick check 1

Hint

You will discover more about the roles of hydrogen bonding in the next few spreads. Remember to look back to the information here to check your understanding.

(a) A water molecule showing the uneven distribution of charge; (b) a cluster of water molecules with hydrogen bonds between them

✔Quick check 2

This table shows the key features of water as a constituent of living organisms.

Property	Key points	Role of water
Good solvent for charged and uncharged substances	Water molecules are attracted to ions and polar molecules, e.g. glucose	Transport in blood, xylem and phloem
Specific heat capacity is high	4.2 kJ are necessary to increase the temperature of 1 kg of water by 1 °C The thermal energy absorbed is used to break hydrogen bonds	Helps prevent changes in body temperature
Latent heat of vaporisation is high	Much thermal energy is used to cause water molecules to change to water vapour – this happens in transpiration in plants, and in sweating and panting in mammals	Coolant – water is used efficiently, as a small amount of water absorbs much thermal energy
High cohesion	Hydrogen bonds 'stick' water molecules together	Helps draw up water in xylem
Can be reactive	Water reacts with other substances	Involved in hydrolysis reactions and in photosynthesis
Incompressibility	Outside pressure cannot force water into a smaller space	Hydrostatic skeleton for some animals e.g. earthworm Provides turgidity in plant cells

Monomers and polymers

The biological molecules covered in the next few spreads are all based on the element carbon. The table shows the four different groups of **macromolecules**, the smaller molecules (sub-units or **monomers**) from which they are formed, and the chemical elements they contain.

Macromolecules	Examples	Sub-unit molecule	Chemical elements
Polysaccharides (Carbohydrates)	Starch, glycogen, cellulose	Glucose	C, H, O
Proteins	Haemoglobin, collagen, enzymes, e.g. amylase, protease, lipase	Amino acids	C, H, O, N, S
Lipids	Triglycerides (fats and oils), phospholipids	Glycerol, fatty acids (and phosphate in phospholipids)	C, H, O (+P, +N for phospholipids)
Nucleic acids	DNA and RNA (mRNA, tRNA and rRNA)	Nucleotides (each nucleotide is composed of a pentose sugar, phosphate and a base)	C, H, O, N, P

✔ *Quick check 3 and 4*

In presenting the table above we have highlighted the fact that you need to know about *four* groups of biological molecules. While lipids are large molecules you will discover on pages 46–47 that they are unlike the other groups because they are not made of repeating sub-unit molecules. Large carbohydrates, such as starch, are made of chains of sugar molecules joined together by covalent bonds. How are lipids different? The sub-unit molecules (glycerol and fatty acids) of triglycerides are different and are not identicial monomers joined end-on to make a chain. So lipids are not polymers like the other three groups.

Sub-unit molecules are joined together by covalent bonds formed in **condensation** reactions that are catalysed by enzymes. These covalent bonds are broken in **hydrolysis** reactions that are catalysed by other enzymes. The next few spreads have examples of these condensation and hydrolysis reactions.

QUICK CHECK QUESTIONS

1 Explain why water is a liquid not a gas like hydrogen sulfide at 40 °C.

2 List the functions of water in living organisms.

3 Name the four groups of macromolecules found in living things, and give an example of each.

4 Make a list of the chemical elements that are found in biological molecules.

Proteins

Module 1

Proteins are macromolecules made of chains of **amino acids**. Twenty different types of amino acid are used to make proteins. Different amino acids have different residual groups (R groups). A single chain of amino acids is a **polypeptide**. Some proteins consist of a single polypeptide; others are composed of two or more.

Key words

- amino acid
- polypeptide
- peptide bond
- condensation
- hydrolysis

Examiner tip

Learn to draw the basic structure of all amino acids. You do not need to know the different R groups. Remember that each type of amino acid has its own specific R group.

✓ *Quick check 1*

Examiner tip

When drawing diagrams like the diagram of peptide bonds here, make sure you show that water is formed from an –OH group and an H.

Amino acids

Cells polymerise amino acids into polypeptides by forming **peptide bonds** (see page 53). A peptide bond forms between an amine group and a carboxylic acid group. When a peptide bond forms, a molecule of water is eliminated (see the diagram of peptide bonds) in a **condensation** reaction. The chemical addition of water breaks the bond and is known as **hydrolysis**.

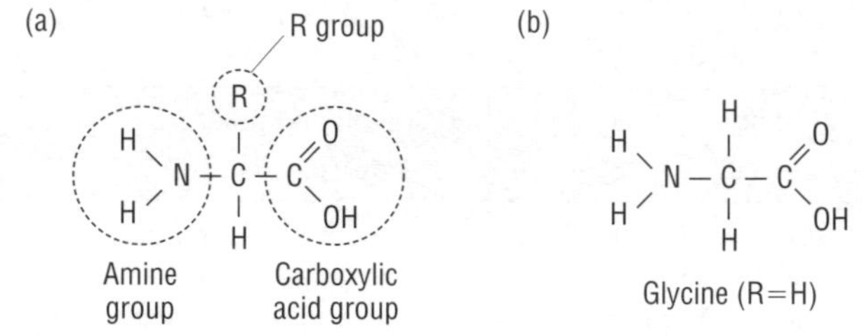

The general structure of all types of amino acid (a) and the smallest, glycine (b)

The properties of proteins depend on the R groups that project from the polypeptide chains. Some are charged (polar) and interact with water. Some are not charged, are hydrophobic, and can interact with phospholipids within membranes. R groups determine the shape of the **active site** of an enzyme (see page 55).

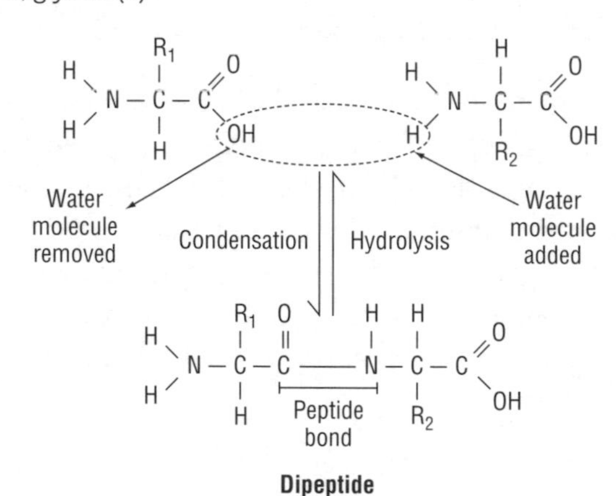

Peptide bonds form by condensation and are broken by hydrolysis. Two amino acids are joined by a peptide bond to form a dipeptide

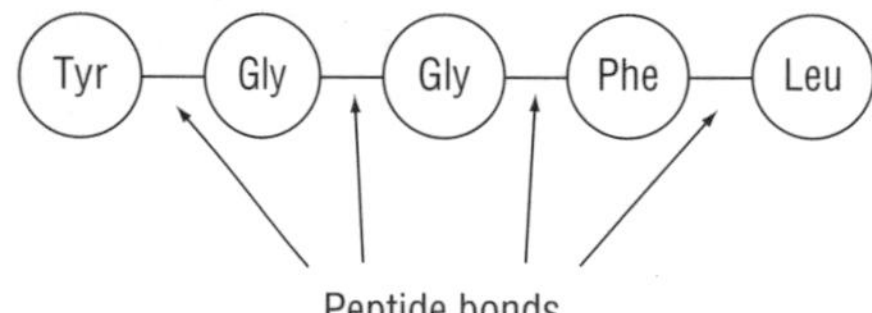

The primary structure of enkephalin, a short peptide made of five amino acids. Polypeptides are made of 10 or more amino acids. The three-letter abbreviations for the amino acids are: tyr, tyrosine; gly, glycine; phe, phenylalanine; leu, leucine.

Organisation of a polypeptide

There are four levels of organisation of a protein. Three levels of structure are described here (for the fourth, quaternary structure, see page 42).

Primary structure

The **primary structure** of a polypeptide is its amino acid sequence. This is determined by the gene that codes for a polypeptide (see page 52).

Secondary structure

Secondary structure is the folding of a polypeptide into one of two structures:

- an α-**helix**, a right-handed helix
- a β-**pleated sheet**, a flat sheet formed by a polypeptide that folds back on itself or links to adjacent polypeptides lying parallel to one another

Parts of the molecule show no regular structure and join α-helices and β-pleated sheets (see 'ribbon' diagram of lysozyme on next page).

Both the helix and the pleated sheet are stabilised by hydrogen bonds.

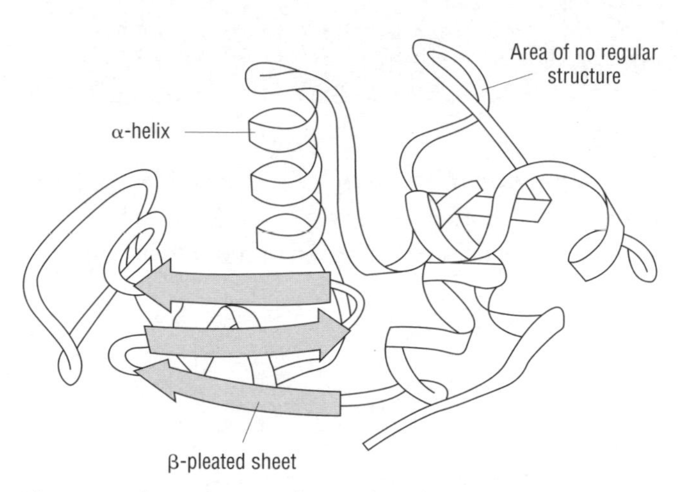

The enzyme lysozyme is a polypeptide with 129 amino acids. This 'ribbon' diagram shows the overall tertiary structure of lysozyme

Tertiary structure

Tertiary structure is the further folding of a polypeptide to give a more complex 3D shape. This shape is very precise and specific to the function of the polypeptide. Some polypeptides, such as the enzyme lysozyme, have areas of their tertiary structure composed of both α-helices and β-pleated sheets. Different parts of the polypeptide chain are close to one another and are stabilised by the following bonds.

- Hydrogen bonds between polar groups anywhere on the polypeptide.
- Disulfide bonds between the sulfur-containing R groups of the amino acid cysteine. Disulfide bonds are covalent bonds, but are susceptible to breakage in the reducing environment of cytosol. They are used to stabilise **extracellular** enzymes, protein hormones (e.g. insulin), lysosomal enzymes, antibodies, and the extracellular portions of transmembrane proteins (e.g. receptor molecules).
- **Ionic bonds** between R groups, which ionise to form positively and negatively charged groups that attract each other.
- Hydrophobic interactions between non-polar R groups.

Hint

Cytosol is the solution of proteins and other substances that surrounds organelles.

In cytoplasm, blood, tissue fluid and other aqueous (watery) fluids, proteins are surrounded by water. Polar R groups interact with water by forming hydrogen bonds that face outwards. Non-polar R groups cannot do this. They tend to cluster towards the centre of the molecule. This gives a hydrophobic core to the molecule that may be useful as the active site of some enzymes.

Some of these bonds break when proteins are heated up or treated with acids and alkalis. When the bonds break, the tertiary structure changes and the protein does not function. This destruction of shape and loss of function is **denaturation**. See page 56, which shows how this affects enzymes. Peptide bonds are very strong and only break by hydrolysis.

✓*Quick check 2, 3, 4*

✓*Quick check 5*

QUICK CHECK QUESTIONS

1 Make an annotated diagram to show how a peptide bond forms between two amino acids.
2 Explain what is meant by primary, secondary and tertiary structure of a polypeptide.
3 Explain how the tertiary structure of a polypeptide is stabilised.
4 Explain the biological significance of the tertiary structure of proteins.
5 Explain the term *denaturation*.

Hint

When answering quick check question 4, it may be a good idea to look at receptors on cell surface membranes (page 7), enzymes (page 54) and antibodies (page 67).

Globular and fibrous proteins

Key words

- haemoglobin
- collagen
- haem
- keratin
- elastin

Many proteins, such as enzymes and **haemoglobin**, are globular and are folded into complex 3D shapes. **Fibrous proteins**, such as **collagen**, have a linear 3D shape and are insoluble. Both haemoglobin and collagen are composed of more than one polypeptide.

Haemoglobin

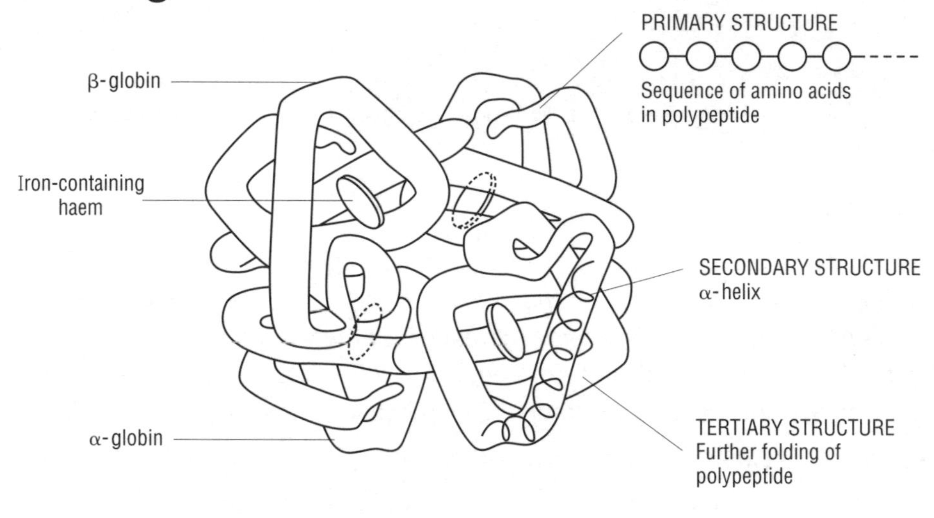

Haemoglobin is a protein showing all four levels of organisation

Hint

The haem group is a non-protein component of haemoglobin.

Examiner tip

Knowing about the structure of haemoglobin will help you to explain how it works – see page 28.

Red blood cells contain many molecules of haemoglobin, a protein that is specialised to transport oxygen. There are four polypeptides in each haemoglobin molecule. These polypeptides are called alpha- and beta-globin (α- and β-globin), and there are two of each. Each polypeptide has a tertiary structure stabilised by hydrophobic interactions in the centre, which maintain the overall shape.

Quick check 1

In the middle of each polypeptide is a **haem** group, which is flat and circular with an atom of iron at the centre. Each haem group combines loosely with one oxygen molecule. This means that one molecule of haemoglobin can carry up to four molecules of oxygen.

Examiner tip

Haemoglobin does *not* have disulfide bonds to stabilise its structure.

Haemoglobin has quaternary structure

A protein has **quaternary structure** if it is made of two or more polypeptides. Haemoglobin has quaternary structure because it has four polypeptides, which fit together and are held in place by interactions between R groups on adjacent polypeptides: these interactions are hydrogen bonds and ionic bonds.

Quick check 2 and 3

Collagen is a fibrous protein

Quick check 4

Collagen is found in skin, bone, cartilage, teeth, tendons, muscles, ligaments and the walls of blood vessels. It gives great toughness to these structures. There are three identical polypeptide chains in a molecule of collagen, wound around each other to form a triple helix. Each polypeptide consists of about 1000 amino acids. In the primary structure, every third amino acid is glycine, which has the smallest R group (one hydrogen atom) of all the amino acids. The sequences of the polypeptides are staggered so that glycine is always found at every position along the triple helix. This allows the three polypeptides to pack closely together to form many hydrogen bonds along their whole length.

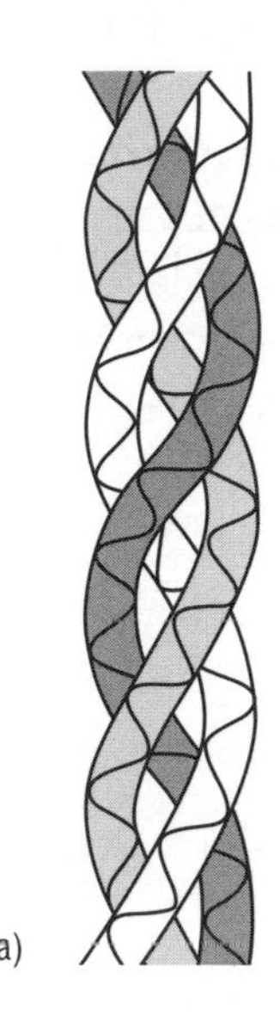

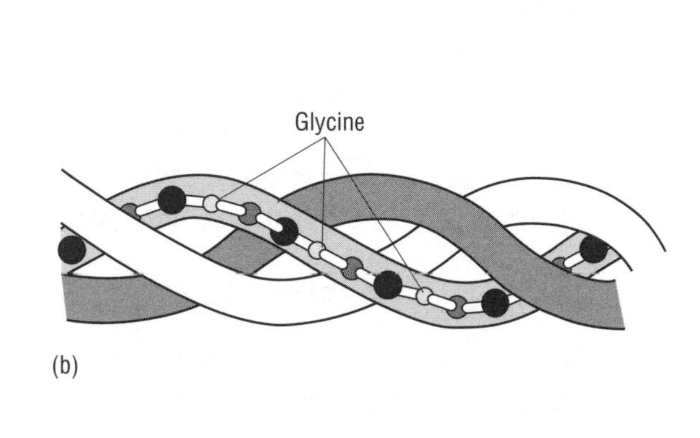

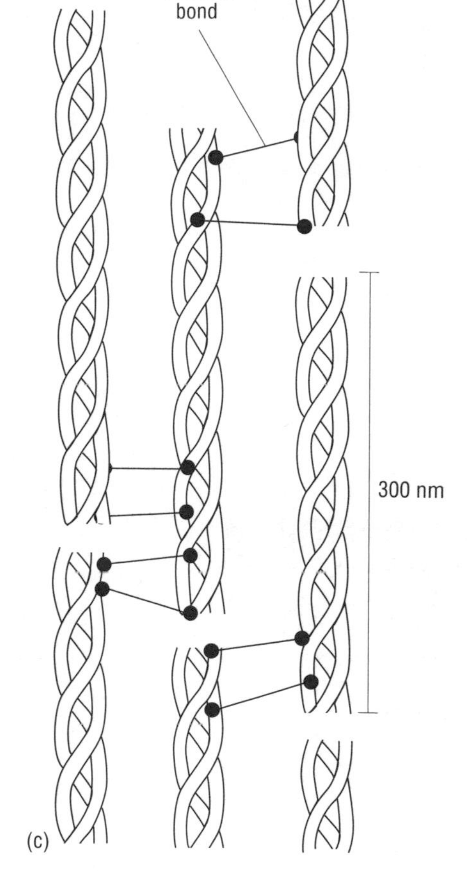

(a) Collagen: three polypeptides are wound very tightly round each other to form a triple helix. (b) A portion of one of the polypeptides showing glycine at every third position. (c) Covalent bonds link the helices together into a strong fibre

Collagen does not show a secondary, tertiary and quaternary structure in the same way as globular proteins such as haemoglobin. The triple helix is a left-handed helix (an α-helix has a right-hand turn), and there is no further folding to give a complex 3D tertiary shape. Adjacent molecules of collagen form covalent bonds between R groups. These link many such molecules into fibres. You can see in diagram (c) above that the molecules are arranged so they overlap without any lines of weakness where the collagen fibre might break if pulled very hard. The many cross-links and the hydrogen bonds within the triple helices give collagen its great strength.

✓Quick check 5 and 6

Other fibrous proteins are **keratin**, which is in skin, nails and hair (see page 64); and **elastin**, which is in blood vessel walls and alveoli (see pages 23 and 19).

QUICK CHECK QUESTIONS

1. Each red blood cell may contain 280 million molecules of haemoglobin. What is the maximum number of oxygen molecules that a red cell can carry?
2. Explain why haemoglobin has a quaternary structure.
3. Suggest why haemoglobin does not have disulfide bonds to stabilise its structure.
4. Define the terms *globular protein* and *fibrous protein*, and give an example of each.
5. Explain how the structure of collagen makes it an ideal substance for ligaments and tendons.
6. Make a table to compare the structure and function of haemoglobin and collagen.

Examiner tip

Collagen is a protein and cellulose is a carbohydrate – but both are very tough. You should be able to compare them in terms of structure and function – see page 45 for details of cellulose.

Carbohydrates

Key words

- carbohydrate
- glycosidic bond
- monosaccharide
- disaccharide
- polysaccharide

Glucose is a small carbohydrate

Carbohydrates are made of the elements carbon, hydrogen and oxygen in the ratio $C_x(H_2O)_y$. Examples are glucose ($C_6H_{12}O_6$) and sucrose ($C_{12}H_{22}O_{11}$). Glucose is a source of energy in organisms, and is built up into macromolecules. It is a **monosaccharide**, which is the simplest form of carbohydrate and cannot be hydrolysed to form other sugars. Glucose is polar so it interacts with water, making it soluble and suitable for transport in the blood. The two forms of glucose (shown in the diagram below) differ only in the position of a hydroxyl group (–OH) about carbon atom 1. This simple difference makes a big difference to the macromolecules formed when they are polymerised. Two other monosaccharides are **fructose** and **galactose**.

Making and breaking a glycosidic bond

When cells make larger molecules from glucose monomers, the chemical bond that forms in a condensation reaction is a **glycosidic bond**. An oxygen atom acts as a 'bridge' between the glucose monomers. **Maltose** is a **disaccharide** in which the oxygen links carbon 1 on one α-glucose monomer with carbon 4 on the other.

Examiner tip

Make sure you can recognise and draw the abbreviated (simple) forms of α- and β-glucose.

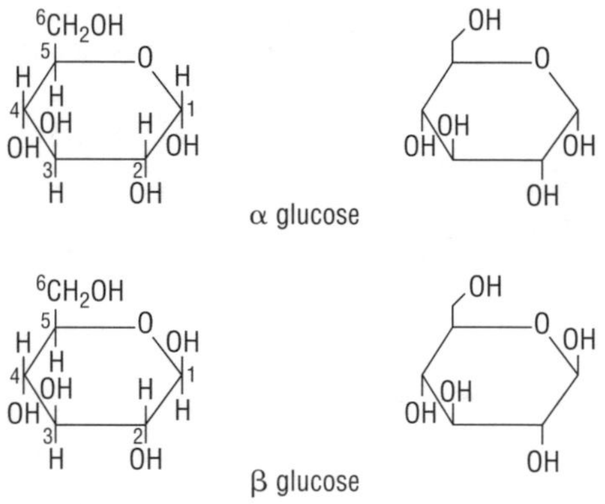

α glucose

β glucose

Alpha (α) and beta (β) glucose molecules in full and abbreviated forms. Notice that the carbon atoms are numbered 1–6, starting next to the oxygen atom and working clockwise

Sucrose is another disaccharide, formed by condensation of the monosaccharides glucose and fructose. Sucrose is the form in which carbohydrate is transported in plants (see page 34). Disaccharides are hydrolysed by extracellular and intracellular enzymes, which catalyse the breakdown of the glycosidic bond by adding water.

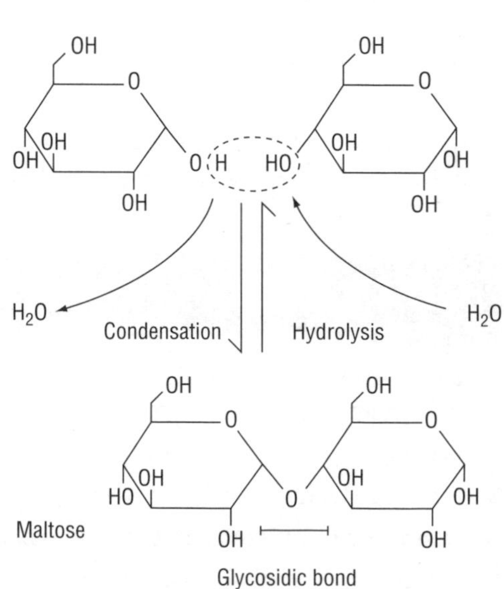

Glycosidic bonds form by condensation and are broken by hydrolysis

Polysaccharides are large carbohydrates

Cells polymerise glucose to form **polysaccharides** that are long-chain molecules with thousands of glucose monomers in each molecule, all joined by glycosidic bonds.

Plants polymerise α-glucose to make two forms of starch:

- **amylose**
- **amylopectin**.

Quick check 1

Quick check 2

Amylose is a single, unbranched **polymer** that forms a helix (spiral shape). An enzyme catalyses the condensation reaction that attaches additional glucose molecules to the end of the chain as shown on the right.

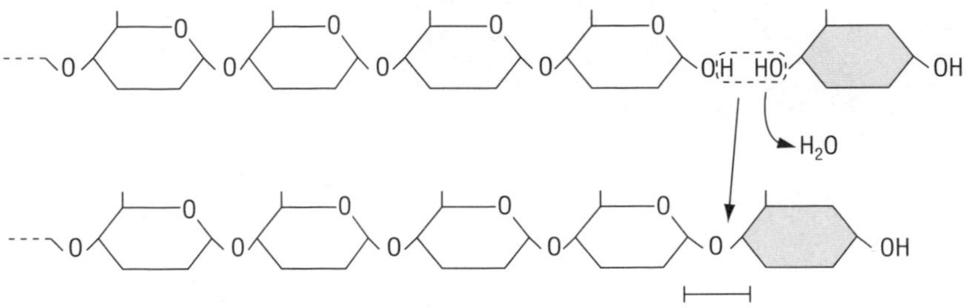

Addition of glucose to a growing end of amylose

Glycosidic bond between C1 and C4

Amylopectin is a branched chain. The branches are made by glycosidic bonds forming between carbon atoms 1 and 6 on glucose. Animals polymerise α-glucose to make **glycogen**, which is like amylopectin except that it branches much more often.

These three molecules (amylose, amylopectin and glycogen) are ideal for energy storage as they are insoluble and compact, and provide large numbers of glucose molecules when needed for respiration. Glycogen and amylopectin have many 'ends' to which glucose can be added or removed as required. This means they are highly efficient as energy-storage molecules. There are different enzymes that catalyse the hydrolysis of these three polysaccharides. One of these enzymes breaks the terminal bonds at the very ends of the molecules, which is the reverse of the reaction for amylose shown in the diagram above. Another removes two glucose units at a time to form maltose.

Quick check 3 and 4

Plants make **cellulose** for cell walls. Cellulose is a polymer of β-glucose. Alternate glucose molecules are turned through 180°. This means that cellulose forms straight chains with many projecting –OH groups that form hydrogen bonds along the molecule, and with adjacent cellulose molecules. This makes cellulose very much stronger than starch and an ideal substance for cell walls, preventing plant cells from bursting when fully turgid. Cellulose molecules are in bundles, known as microfibrils, and these are laid down in cell walls in different directions to give added strength to the cell walls.

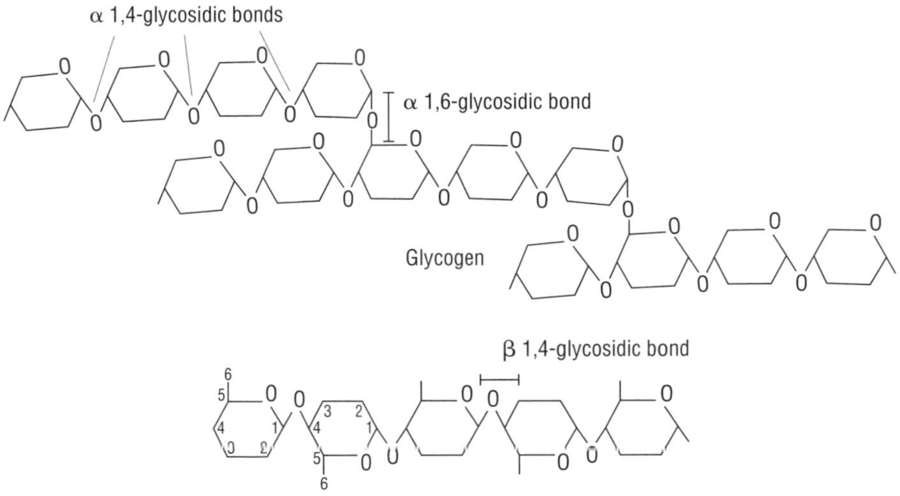

Glycogen and cellulose

Quick check 5 and 6

QUICK CHECK QUESTIONS

1. Define the following terms: *monosaccharide, disaccharide, polysaccharide, glycosidic bond.*
2. Describe how a glycosidic bond between two molecules of α-glucose is formed, and how it is broken.
3. Explain why starch and glycogen are good stores of energy.
4. Draw a diagram to show how a maltose molecule is formed from the end of amylose. Name the enzyme that catalyses this reaction.
5. Make a table to compare the polysaccharides mentioned in this spread.
6. Explain why cellulose is an ideal molecule for plant cell walls.

Lipids

Key words

- lipid
- triglyceride
- cholesterol
- saturated
- unsaturated

Lipids are large molecules with few oxygen atoms and many carbon and hydrogen atoms. This chemical composition makes them water repellent, or hydrophobic. They are less dense than water. Two important groups of lipids are **triglycerides** (fats and oils) and **phospholipids**. These are made from sub-unit molecules – **glycerol** and **fatty acids**. Glycerol is a three-carbon compound. Each carbon atom has a hydroxyl group (–OH group), which can react with a fatty acid to form an **ester bond**.

Fatty acids are long-chain hydrocarbon molecules. Each fatty acid has a carboxyl group at one end that can react with an –OH group on glycerol. Fatty acids vary as follows:

- length of the hydrocarbon chain (12–20 carbon atoms)
- number of double bonds between carbon atoms in the chain.

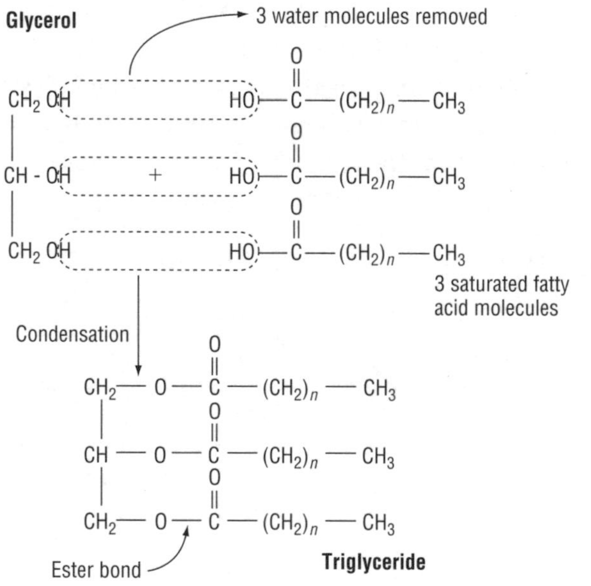

Glycerol

3 water molecules removed

CH_2OH + $HO-C(=O)-(CH_2)_n-CH_3$

$CH-OH$ + $HO-C(=O)-(CH_2)_n-CH_3$

CH_2OH + $HO-C(=O)-(CH_2)_n-CH_3$

3 saturated fatty acid molecules

Condensation

$CH_2-O-C(=O)-(CH_2)_n-CH_3$

$CH-O-C(=O)-(CH_2)_n-CH_3$

$CH_2-O-C(=O)-(CH_2)_n-CH_3$

Ester bond

Triglyceride

n = 12 to 20

A triglyceride is formed when three fatty acids form ester bonds with glycerol. This is another example of a condensation reaction

Quick check 1

Saturated and unsaturated fatty acids

Saturated fatty acids have no double bonds in the hydrocarbon chain. Unsaturated fatty acids have one or more double bonds somewhere along the chain. Monounsaturated fatty acids have one; polyunsaturated fatty acids have two or more double bonds in the chain. Triglycerides and phospholipids always have a mixture of different fatty acids. Triglycerides from fish and plants are oils, as they contain more unsaturated than saturated fatty acids. An oil is a liquid at room temperature. Triglycerides from mammals are fats, as they contain more saturated fatty acids. A fat is solid at room temperature.

Quick check 2

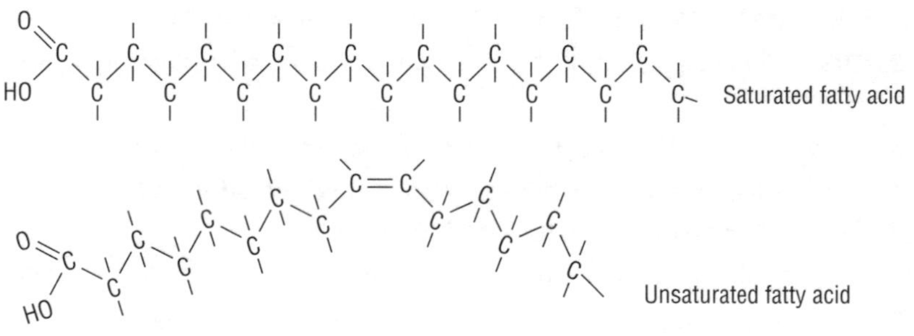

Saturated fatty acid

Unsaturated fatty acid

The hydrocarbon chains of a saturated and an unsaturated fatty acid. All the hydrogen atoms from the hydrocarbon chain have been omitted. The double bond causes the chain to bend

Functions of triglycerides

Animals and plants use triglycerides for:

- energy stores – when they are respired, fats and oils release much energy (39 $kJ\,g^{-1}$)
- thermal insulators – important to mammals and birds that live in cold climates (e.g. marine mammals such as seals and whales)
- buoyancy – fats and oils are less dense than water
- protecting internal organs, such as kidneys
- an important source of water (especially for desert animals) when respired.

Phospholipids

The diagram opposite shows two ways in which we can show phospholipid molecules. The jagged lines on (a) and the straight lines on (b) are different ways of showing the hydrocarbon chains.

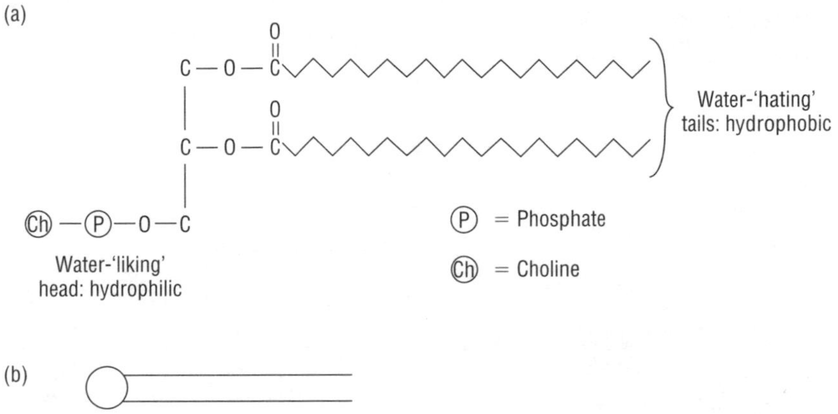

(a)

Water-'hating' tails: hydrophobic

Water-'liking' head: hydrophilic

Ⓟ = Phosphate

Ⓒⓗ = Choline

(b)

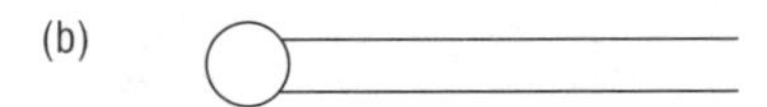

(a) Phospholipids differ from triglycerides by having a phosphate group attached to glycerol in the place of a fatty acid; (b) a simple way of drawing a phospholipid

Attached to the phosphate are other groups (e.g. choline), which dissolve in water. The fatty acids are hydrophobic, repelling water. Phospholipids therefore form stable bilayers when in water (see the figures on page 6). These have hydrophobic cores and hydrophilic surfaces. The bilayer is an ideal structure for cell membranes, as it creates a hydrophobic barrier between the parts of a cell, or between a cell and its surroundings.

✔Quick check 3

✔Quick check 4

Cholesterol

Cholesterol is a lipid. The –OH group makes it polar at one end, while the four hydrocarbon rings and the hydrocarbon tail are non-polar. it is arranged in bilayers in a similar way to phospholipids, with which it interacts to control membrane fluidity (see page 7). All **steroids**, such as testosterone and progesterone, have the four-ring structure and are made from cholesterol.

Hint

There is more about cholesterol on page 61.

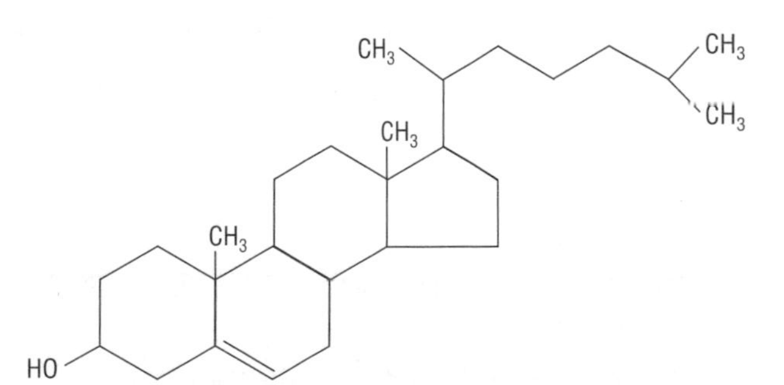

The four-ring structure of cholesterol

✔Quick check 5

1. Describe the structure of a triglyceride molecule.
2. Describe how triglycerides may differ from one another.
3. Explain how a phospholipid differs from a triglyceride.
4. Explain how phospholipids form bilayers.
5. Outline the importance of cholesterol to human metabolism.

Practical work with biological molecules

Testing for biological molecules

Test for starch

The iodine test for starch:

- test substance may be solid or liquid
- place a small amount on a white tile, or put it in a test tube
- add iodine solution (this is a yellow-orange colour)
- positive result – a blue-black colour
- negative result – a yellow-orange colour (iodine solution does not change colour).

Iodine binds to the centre of the amylose helix, changing its colour from yellow to blue–black.

Test for reducing sugars

Benedict's test for reducing sugars:

- place about 1 cm^3 of a solution of the test substance in a test tube
- add the same volume of Benedict's solution (this is blue)
- boil or heat to 80 °C in a water bath
- positive result – a colour change from blue to green, then yellow/orange, then red with a precipitate (the final colour in Benedict's test depends on the concentration of reducing sugar in the test substance – if it is very low, the final colour will be green)
- negative result – no change to the blue colour.

Almost all sugars are reducing agents and are known as **reducing** sugars. When they are mixed with an oxidising agent and heated, a redox reaction takes place. **Benedict's solution** contains copper sulfate with divalent copper ions (Cu^{2+}). When they are reacted with a reducing sugar these ions change to Cu^{+}, which forms a precipitate of red copper(I) oxide. This is the chemical basis of **Benedict's test**.

Benedict's test does not distinguish between different reducing sugars (e.g. maltose and glucose). Most sugars are reducing sugars – but sucrose is not.

Test for non-reducing sugars

Sucrose is the only common non-reducing sugar. If a test substance gives a *negative result* with Benedict's test:

- add several drops of dilute hydrochloric acid to 2 cm^3 of the test solution in a test tube and boil in a water bath for a few minutes (the acid acts as a catalyst to hydrolyse sucrose to glucose and fructose)
- cool the test tube and add an alkali *before* doing Benedict's test – this neutralises the acid (acid solutions inhibit Benedict's test)
- carry out Benedict's test as described above
- if the result is positive, you know that non-reducing sugars (probably sucrose) were present in the original test substance, but there were no reducing sugars
- If the result is negative (no colour change), you know that there were no non-reducing sugars and no reducing sugars.

Key words

- iodine solution
- Benedict's solution
- biuret test
- emulsion test
- colorimeter

Examiner tip

For each test, make sure you know the chemical reagent(s) to use, the procedure to follow, and the positive and negative results.

Examiner tip

The full name is 'iodine in potassium iodide solution', but it is usually known as 'iodine solution'. Do *not* call it 'iodine' – that is the element.

✓*Quick check 1*

✓*Quick check 2*

Test for proteins

The **biuret test** for proteins:

- place about 1 cm^3 of a solution or suspension of the test substance in a test tube
- add an equal volume (about 1 cm^3) of biuret solution (copper sulfate and sodium hydroxide) and mix by swirling the tube
- positive result – a violet/purple/lilac colour
- negative result – no change to the blue colour of copper sulfate.

Test for lipids

The **ethanol emulsion test** for lipids:

- crush the material to be tested with a glass rod in some ethanol in a test tube
- filter, or carefully decant, the ethanol into a second test tube with cold water
- discard the solid residue – avoid any solid getting into the water
- do *not* mix the ethanol and the cold water – the ethanol will float on the water
- positive result – a milky emulsion forming in the water indicates lipid in the ethanol.

Using a colorimeter

To make the Benedict's a *quantitative* test, we can use a **colorimeter** to give us readings. The colorimeter measures the degree of 'blueness' of the solution remaining after the reducing sugar has reacted with Benedict's solution. After boiling a small sample of the test solution with a larger volume of Benedict's solution, filter the test solution (or spin it in a centrifuge) to remove the precipitate. A solution that remains blue has no reducing sugar in it and gives a high reading for absorbance in the colorimeter. A solution that has no blue colour has a high concentration of reducing sugar and gives a low reading for absorbance.

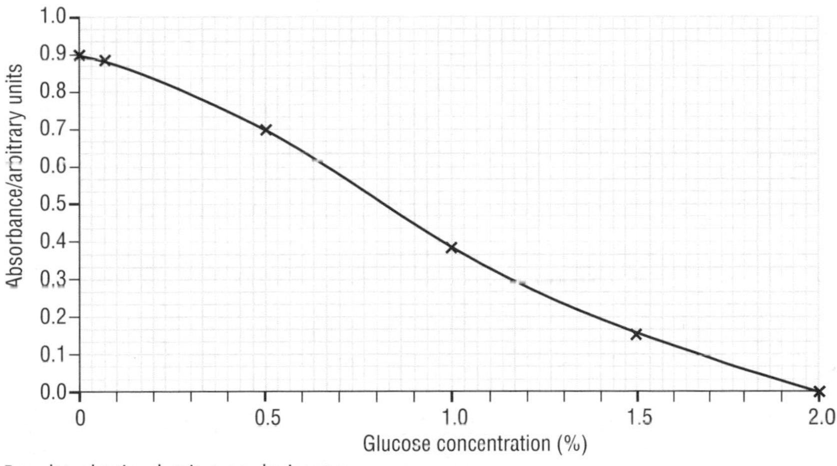

Results obtained using a colorimeter

Hint

There is more about colorimetry on page 59.

Quick check 3 and 4

1 State why Benedict's test does not distinguish between glucose and maltose.

2 Plant material often contains reducing sugars and non-reducing sugars. How would you use Benedict's test to confirm this?

3 A student tested some plant material with Benedict's test and, using a colorimeter, found the absorbance of the solution to be 0.6 arbitrary units. Use the graph above to find the concentration of reducing sugar.

4 Colorimetry works well with the biuret test. Describe how you would use colorimetry to find the concentration of protein in seeds, for example broad beans.

Nucleic acids – DNA and RNA

Key words

- nucleic acid
- deoxyribonucleic acid
- ribonucleic acid
- nucleotide
- polynucleotide
- purine
- pyrimidine

The **nucleic acids** – deoxyribonucleic acid (**DNA**) and ribonucleic acid (**RNA**) – make up the cell's information storage and retrieval system. DNA is a long-term store of genetic information that is passed from cell to cell during growth, and from parent to offspring during reproduction. RNA molecules are smaller and do not last as long as DNA. They have a number of different functions in information retrieval when proteins are synthesised in cells.

Nucleic acids are polynucleotides

DNA and RNA are macromolecules made of chains of **nucleotides**. These chains of nucleotides are called **polynucleotides**. Each nucleotide consists of:

- pentose sugar (with five carbon atoms)
- phosphate
- nitrogen-containing (nitrogenous) base.

There are two different types of base.

- The larger bases are the **purines** – **adenine** and **guanine**. They have double rings of carbon and nitrogen atoms.
- The smaller bases are the **pyrimidines** – **thymine**, **cytosine** and **uracil**. These have a single ring of carbon and nitrogen atoms.

Examiner tip

DNA and RNA are made of nucleotides – *not* amino acids.

The bases are often known by the letters A, G, T, C and U:

- DNA is made up of A, G, C and T
- RNA is made up of A, G, C and U.

Examiner tip

Remember – uracil replaces thymine in RNA.

A nucleotide is shown in the diagram below. DNA nucleotides have the sugar **deoxyribose**, RNA nucleotides have the sugar **ribose**. Both sugars are pentoses – they have five carbon atoms. The numbers refer to the carbon atoms. Note that this is a schematic diagram, and many of the groups (such as –OH) have been left out.

The nucleotides are attached to each other by covalent bonds between phosphate and sugar molecules to form a 'backbone'.

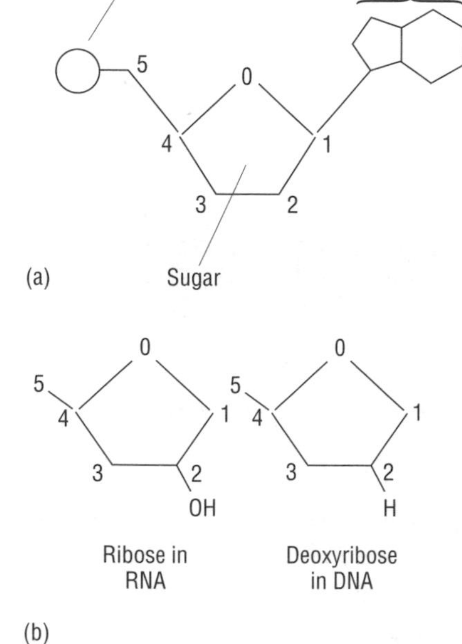

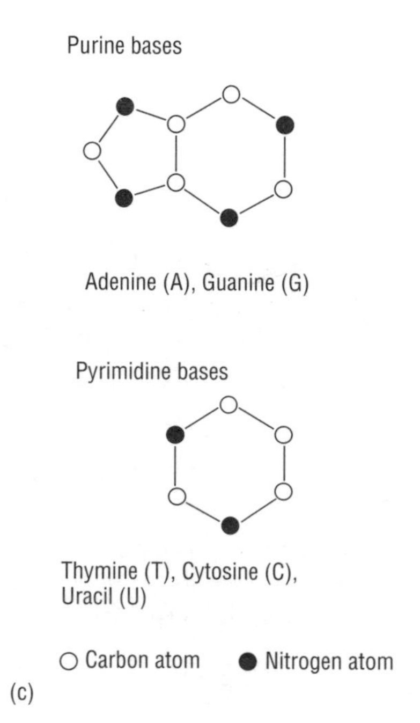

(a) A nucleotide; (b) ribose and deoxyribose; (c) simplified structure of purine and pyrimidine bases. The positions of the carbon atoms in the sugars are numbered 1 to 5.

DNA is a double helix

The figure opposite shows the **double helix** of DNA. It is made of two polynucleotides held together by hydrogen bonds. The bases are shown as blocks projecting inwards; the dashed lines represent hydrogen bonds.

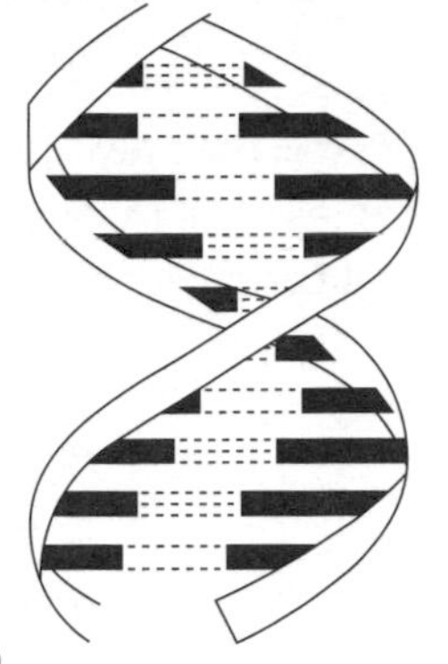

The bases are always arranged so that a purine is opposite a pyrimidine: A with T; G with C. There are two hydrogen bonds between A and T, and three between G and C. The bases are complementary in size and shape, so that only the pairings A–T and G–C fit into the space between the sugar-phosphate backbone of DNA. The two polynucleotides are antiparallel. The carbon atoms in deoxyribose in the diagram opposite are numbered 1–5. The sugars are arranged with carbon 3 pointing downwards on one side and upwards on the other, so that one polynucleotide is in the 3′ (3 prime) to 5′ (5 prime) direction, and the other in the 5′ to 3′ direction. Each polynucleotide assumes a helical shape as the base pairs are not at a right angle to the sugar-phosphate backbone – hence the double helix.

DNA is well suited for the long-term storage of genetic information as:

- it is a very stable molecule, so the information stored in DNA is kept for a long time
- it is a large molecule that stores huge amounts of information
- each of the two polynucleotides acts as a template for synthesis of a new polynucleotide during DNA replication (see page 52).

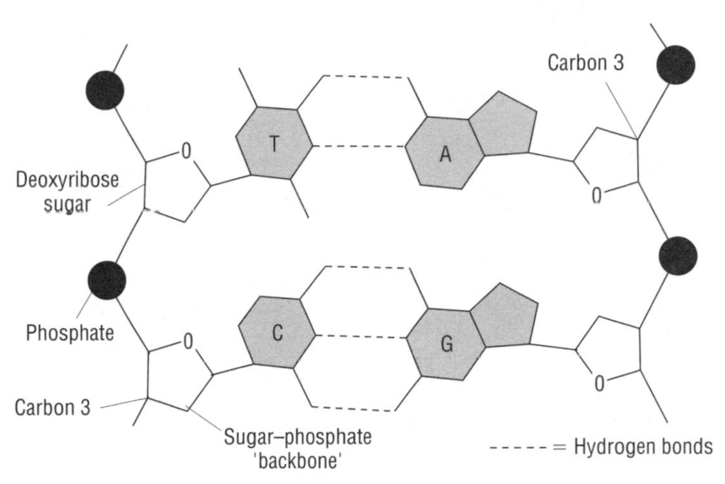

This depiction of a small part of a DNA molecule shows that the polynucleotides are antiparallel – look carefully at the deoxyribose molecules

Examiner tip

The **base-pairing rules** (A–T and C–G) always apply. Look out for more examples in the next few spreads.

Hint

If you need to refresh your memory about hydrogen bonds, see page 38.

Three forms of RNA

RNA differs from DNA in that the molecule is a single-stranded polynucleotide and is often shorter than DNA. RNA contains the sugar ribose, not deoxyribose, and the base uracil, not thymine. The table shows the three forms of RNA and their functions.

Quick check 1 and 2

Type of RNA	Structure of polynucleotide	Function
Messenger RNA (mRNA)	Variable length, no base pairing	Transfers genetic information from DNA to ribosomes, after which it is broken down
Transfer RNA (tRNA)	Clover-leaf structure with some base pairing	Carries amino acids to ribosomes
Ribosomal RNA (rRNA)	Folded and attached to proteins to make ribosomes	Provides site for assembly of amino acids to make proteins

Quick check 3 and 4

QUICK CHECK QUESTIONS

1. Describe the structure of a DNA molecule.
2. Explain what is meant by base pairing in DNA.
3. Make a table to show how the structure of DNA differs from the structure of RNA.
4. Describe the functions of DNA and RNA.

Hint

In quick check question 4, remember there are *three* different types of RNA.

Replication and roles of DNA

Module 1

DNA is the only molecule that can be copied. There are enzymes in cells that make copies of DNA, but they need some DNA to start with. This copying process, or replication, is very precise. There are very few copying errors.

Key words

- replication
- semi-conservative
- transcription
- amino acid activation
- translation

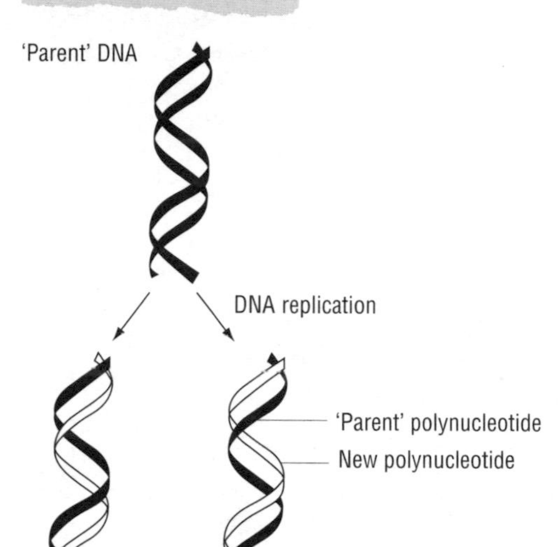

The principle of semi-conservative replication

Semi-conservative replication

Each polynucleotide acts as a template for making a new polynucleotide. This means that a new molecule of DNA consists of an older polynucleotide and a recently made one.

This type of replication is called **semi-conservative replication**, as half of the 'parent' DNA molecule is passed to the 'daughter' molecule. In this way the parent DNA is conserved as one polynucleotide forms half of one daughter molecule and the other polynucleotide forms part of the other one. You can see this in the diagram opposite.

Replication in more detail

1. The double helix unwinds and the DNA 'unzips' as hydrogen bonds between the polynucleotides are broken.
2. Existing polynucleotides act as templates for assembly of nucleotides.
3. Free nucleotides, which have been made in the cytoplasm, move towards the exposed bases of DNA.
4. Base pairing occurs between free nucleotides and exposed bases. A matches T and C matches G. Hydrogen bonds form between complementary bases.
5. The enzyme DNA polymerase forms covalent bonds between the free nucleotides attached to each template.
6. Two daughter DNA molecules form separate double helices.

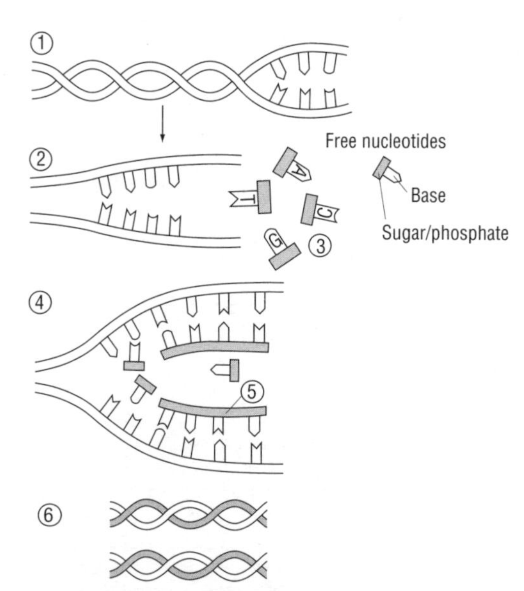

What happens to a small section of DNA when it is replicated

Replication in eukaryotes occurs during the S phase of the cell cycle (see the diagram of the mitotic cell cycle on page 13). It is a highly controlled process that involves several enzymes. The main enzyme is DNA polymerase, which matches the free nucleotides with the exposed bases on the template polynucleotide. It builds the new polynucleotide in the 5′ to 3′ direction, and as it goes it forms covalent bonds between the nucleotides on the new polynucleotide. Occasionally, mistakes are made and the wrong nucleotide is matched with the exposed base on the template strand. DNA polymerase is also a proofreading enzyme. When a mistake happens, the 'wrong' nucleotide is cut out and the correct one inserted.

Quick check 1

Hint

See the diagram of a small part of a DNA molecule on page 51 to remind yourself about the 5′ to 3′ direction.

The functions of DNA

DNA is a long-term store of genetic information. It is a very stable molecule and can last for a very long time. Aside from being a store, DNA codes for messenger RNA, which is used by ribosomes to assemble amino acids to make polypeptides, transfer RNA and ribosomal RNA.

Protein synthesis

There is a huge amount of information stored in the DNA of our chromosomes. Each cell has a full complement of chromosomes with all the genetic information needed to make the whole body. However, individual cells need to retrieve only a small amount of this information.

For example, cells in our salivary glands use the gene that codes for the enzyme amylase. Liver cells use the gene that codes for catalase. Each gene is a code for the primary structure of a polypeptide: the sequence of bases in DNA determines the sequence of amino acids in the polypeptide, which is made by ribosomes in the cytoplasm. Cells have thousands of ribosomes that use short-lived transcripts of the genes in the form of messenger RNA (mRNA). The process of protein synthesis occurs in four stages.

Hint

A gene is a length of DNA that codes for a polypeptide.

Hint

See End-of-unit questions for Unit 1 (question 1, page 36) for a diagram of the organelles involved in protein synthesis.

Transcription

DNA 'unzips' along the length of the gene, so the enzyme RNA polymerase can match free RNA nucleotides to form a molecule that is complementary to the template polynucleotide. This follows the rules of base pairing. The other polynucleotide takes no part in this process. Each group of three bases (or triplet) in mRNA codes for a specific amino acid.

Hint

mRNA is made by assembling nucleotides when a transcript of a gene is needed.

✓*Quick check 2*

Movement of mRNA to ribosomes

When transcription is finished, the messenger RNA molecule moves from the chromosome to a ribosome in the cytoplasm. In a eukaryotic cell, mRNA moves from the nucleus, where transcription occurs, through nuclear pores to ribosomes.

Amino acid activation

Enzymes attach amino acids to their specific tRNA molecule. This needs energy supplied by ATP.

Translation

The mRNA molecule binds to a ribosome, and translation begins. There are two places on a ribosome where tRNA molecules, with their amino acids, can attach. The first two tRNA molecules occupy these two sites and a peptide bond forms between the amino acids. The ribosome 'reads' the sequence of bases on mRNA in groups of three (triplets). The tRNA molecules identify the amino acids, so the ribosome puts the amino acids together in the sequence dictated by the triplets of mRNA. When a ribosome has translated the mRNA, a polypeptide is released that either goes into the cytosol to be used, or goes via the space inside rough endoplasmic reticulum to the Golgi apparatus to be modified, packaged and put into a **vesicle**.

Examiner tip

You need to know the main stages of protein synthesis – the details are part of the A2 course.

✓*Quick check 3*

QUICK CHECK QUESTIONS

1 Describe what happens when DNA is replicated.

2 Describe the role of DNA in protein synthesis.

3 Outline the roles of the nucleus, DNA, mRNA, ribosomes, tRNA, rough endoplasmic reticulum and Golgi apparatus in the synthesis of a polypeptide.

Hint

In quick check question 3, *outline* means you should include the main points as given in this spread; you do not have to give any further details.

Enzymes

Key words

- intracellular
- extracellular
- activation energy
- specificity
- active site

Many chemical reactions occur in living organisms. The sum of all the chemical reactions that occur in the body is known as **metabolism**. Most of these reactions would occur only very slowly if we did not have special **catalysts**. **Enzymes** are biological catalysts made of protein. Enzymes convert substrates into products by having **active sites**, where reactions occur. Some enzymes speed up reactions where molecules are broken down (catabolic reactions); others catalyse reactions where large molecules are built up from smaller ones (anabolic reactions).

Hint

Enzymes catalyse hydrolysis and condensation reactions (see pages 40, 44 and 46 for examples).

Quick check 1

Intracellular enzymes

Intracellular enzymes work inside cells, most catalysing reactions that occur in series. These are multi-step processes, such as respiration and photosynthesis, which are studied in detail in the A2 course. Some intracellular reactions are condensation reactions, such as those described in earlier spreads in this module.

Extracellular enzymes

These work outside cells, catalysing hydrolysis reactions to break down macromolecules into small, soluble molecules that can be absorbed (see page 39).

This table summarises the properties of enzymes.

Protein properties	Catalyst properties
Globular proteins Tertiary structure has specific 3D shape which provides active site where reaction occurs Influenced by temperature and pH Denatured by extremes of pH and temperatures above optimum	Increase rates of reaction Specific to a reaction Remain unchanged at end of reaction Lower the activation energy

Reactions will occur more quickly if heated. As the structure of many proteins is permanently changed at temperatures greater than 40 °C, this is not an option for living organisms.

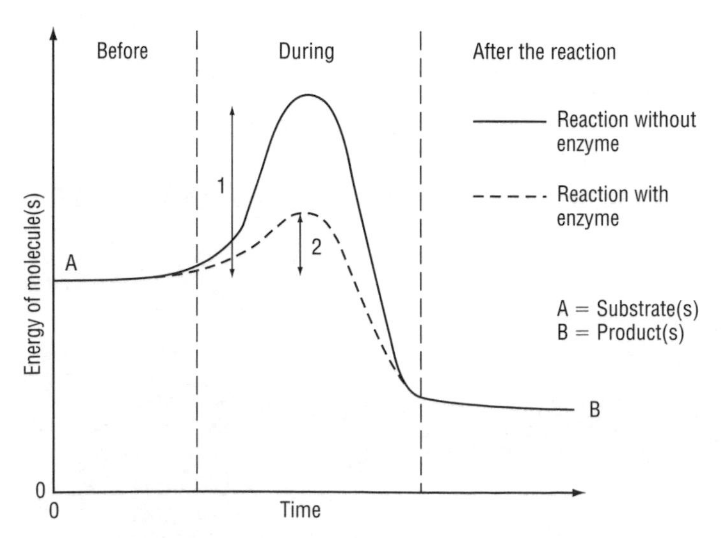

An enzyme lowers activation energy for this exergonic reaction, in which energy is released from a substrate. Activation energy 1, without enzyme; 2, with enzyme

Activation energy

Energy is stored in the bonds that hold atoms together in a molecule. When substrates react, bonds are broken and new bonds form. When there is no enzyme, reactions do not occur readily because the substrate (or substrates) do not have enough energy to be converted to a product. The extra energy that needs to be given is the activation energy. Enzymes decrease activation energy by providing an **active site** where reactions occur more easily than elsewhere.

Enzymes allow covalent bonds in substrate molecules to be broken and energy to be released. In exergonic reactions, energy is often transferred to ATP. This happens in respiration.

Action of enzymes

In a solution, enzyme and substrate molecules are constantly moving and frequently collide. A substrate molecule may fit into the enzyme active site. It is held in the active site for a brief moment, forming an **enzyme–substrate complex**.

The reaction occurs and the product or products are formed, giving an **enzyme–product complex** that dissociates to release the product(s) from the active site. The active site is now free to receive another substrate molecule.

Substrate molecules have a shape that is complementary to the shape of the active site – the substrate fits into the active site like a key fitting into a lock. This idea is the **lock-and-key hypothesis**.

The diagram opposite shows how the enzyme changes shape to mould itself around the substrate molecule. This is called **induced fit**. R groups on the polypeptide forming the active site move to form temporary bonds with the substrate. Two substrate molecules may be brought close together so that they react. The bonds within a single substrate may be put under strain so that it is broken down into two products.

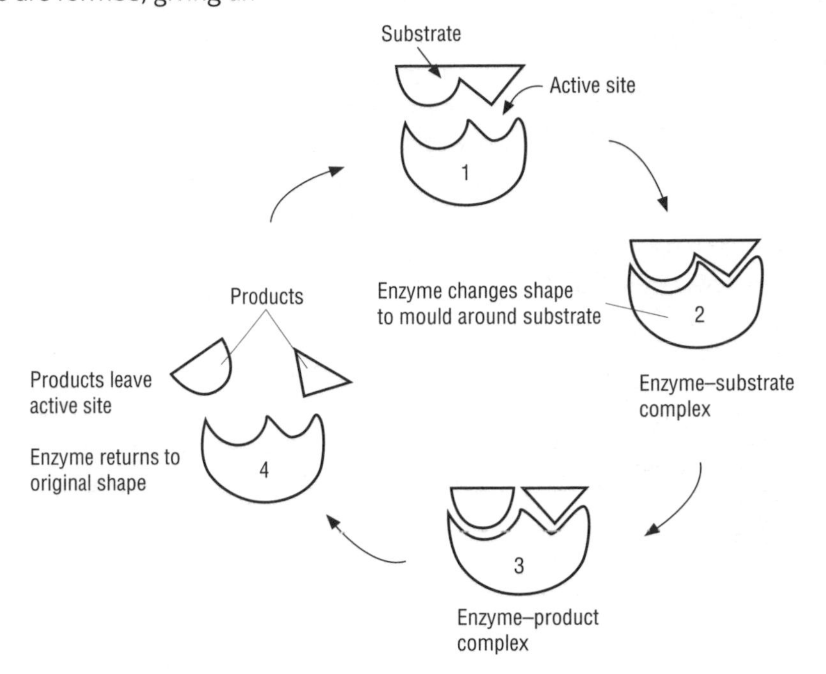

Induced fit: enzyme molecules change shape as they bind to their substrates

Examiner tip

The active site is part of the enzyme, not part of the substrate.

Module 1

✓ Quick check 2 and 3

Enzyme specificity

An active site is specific for one type of molecule. Some examples of enzyme specificity are seen in the following extracellular enzymes:

- amylase – hydrolyses glycosidic bonds in starch to form maltose
- subtilisin – hydrolyses peptide bonds between any pair of amino acids
- trypsin – hydrolyses only peptide bonds next to arginine and lysine.

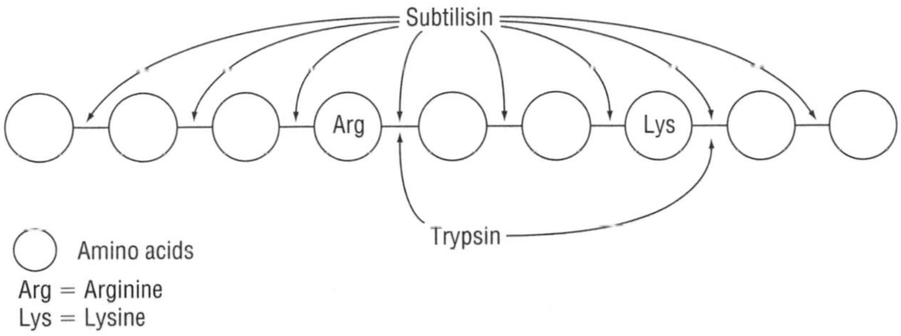

Subtilisin

Arg Lys

Trypsin

Amino acids
Arg = Arginine
Lys = Lysine

The peptide bonds hydrolysed by the enzymes subtilisin and trypsin. Both are proteases

Examiner tip

This is all to do with shape again. Amylase will not break down cellulose because only starch molecules have the right shape to fit in the active site.

✓ Quick check 4

QUICK CHECK QUESTIONS

1 What happens in catabolic and anabolic reactions?

2 Explain the difference between the lock-and-key hypothesis and the induced-fit hypothesis of enzyme action.

3 Explain how an enzyme increases the rate of a reaction.

4 Make a diagram, similar to the diagram of induced fit above, to show two substrate molecules joining together to form one product molecule.

Factors affecting enzyme activity

Key words

- optimum
- limiting factor
- inhibitor
- cofactor
- coenzyme

Examiner tip

Not all enzymes are denatured at a temperature of 40–50 °C. Enzymes from bacteria that live in hot springs are still active at temperatures of 70 °C or higher.

✓*Quick check 1*

Temperature

At **M** on the figure opposite, there is a slow reaction because molecules of the enzyme and substrate have little kinetic energy and rarely collide. At **N**, there is more kinetic energy and collisions occur more frequently. **O** is the **optimum** temperature, where the rate is fastest. **P** is where the enzyme molecules begin to lose their tertiary structure. Bonds that hold the polypeptides in specific shapes are broken and the shape of the active site changes. Substrate molecules no longer fit. At **Q**, the enzyme is denatured and loses all activity.

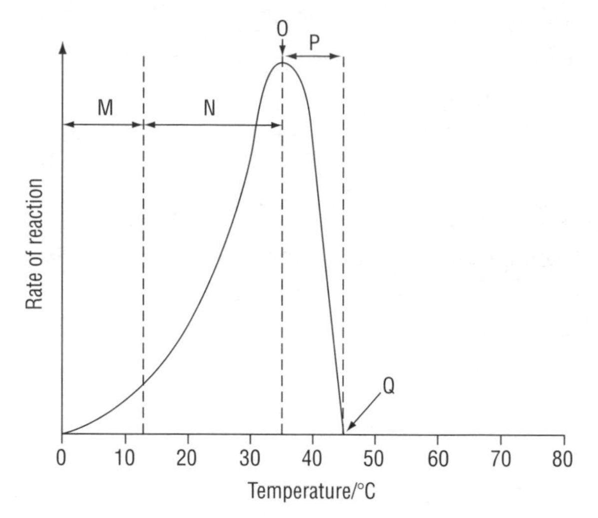

Not all enzymes have an optimum temperature around 40 °C. For enzymes from some plants, the optimum is nearer 20 °C. Enzymes in animals that live in very cold environments may have optimum temperatures in the range 5–10 °C.

Substrate concentration

Each point plotted is the rate of reaction with different starting concentrations of substrate, but the same concentration of enzyme has been used each time. In the figure opposite, in **A** the rate increases as the substrate concentration increases. Substrate concentration is the **limiting factor**. In **B** the rate does not increase any further as the enzyme concentration is limiting. All the active sites are filled.

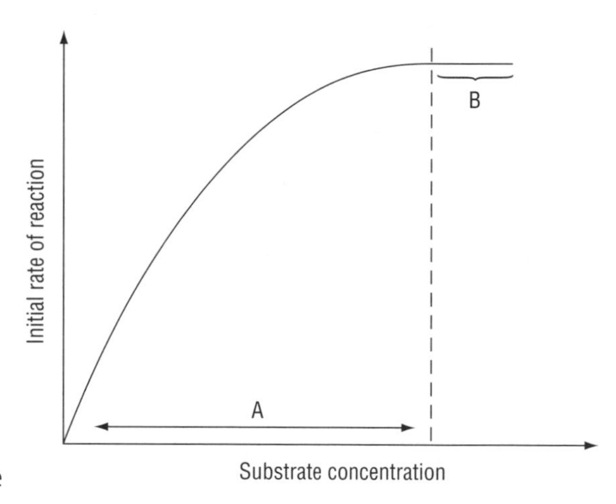

Examiner tip

Note how these graphs have been divided into sections. This is a good technique to use when answering questions.

Enzyme concentration

The rate of reaction increases as the concentration of enzyme increases. When there is plenty of substrate, the only limiting factor is the enzyme concentration. With more enzyme molecules, there are more active sites available.

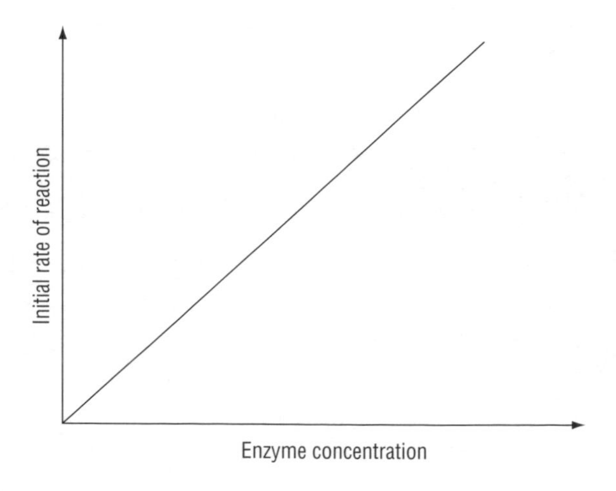

pH

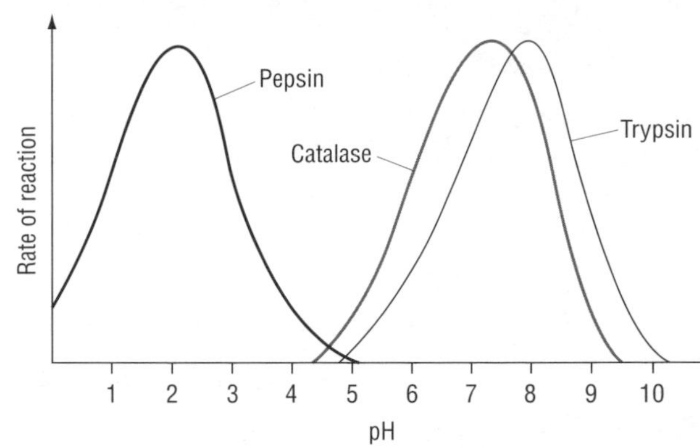

Most enzymes work over a narrow pH range. The figure opposite shows the optimum pH for three different enzymes. The optimum pH for intracellular enzymes, such as catalase, is about 7.0. Pepsin works in the stomach, where hydrochloric acid is secreted. This explains its low optimum pH. Trypsin works in the small intestine, which has a pH of about 8.0.

When the pH changes from the optimum:

- the shape of the enzyme changes
- the affinity of the substrate for the active site decreases.

Many of the bonds that hold an enzyme's tertiary structure together are bonds between positively and negatively charged R groups. When the pH changes, there is a change in the concentration of hydrogen ions that cancels out the charge on these R groups so the tertiary structure changes, which means that active sites lose their specific shape and substrate molecules do not fit.

✓Quick check 2

Some enzymes are made only of protein – they may consist of one polypeptide, or several. They combine with their substrate(s), catalyse their reaction, release their product(s), and start again. Most enzymes are not like this – they need ions or organic molecules to function. Enzymes can also be inhibited. Many substances that are poisonous to us are enzyme **inhibitors**; many medicinal drugs are also enzyme inhibitors.

Substance	Definition	Examples
Cofactor	A non-protein component of an enzyme – inorganic or organic – required by enzymes to carry out reactions	• Metal ions, e.g. zinc ions (Zn^{2+}) in carbonic anhydrase • Small organic molecules, e.g. haem in catalase
Coenzyme	An organic cofactor. Some coenzymes are tightly bound to their enzyme (e.g. FAD), others are mobile within the cell (e.g. NAD). Many are involved in the energy-transfer reactions of respiration and photosynthesis	• NAD, FAD and coenzyme A – involved in reactions in respiration • NADP – involved in reactions in photosynthesis
Competitive inhibitor*	Molecular shape is similar to substrate, so fits into active site, *slowing* rate of reaction	• Statins – compete with cholesterol for active site of HMG–CoA reductase, a liver enzyme which helps to make cholesterol (reversible) • Penicillin (an antibiotic) – inhibits an enzyme that makes cell walls in some bacteria (non-reversible)
Non-competitive inhibitor*	Molecule binds to a part of the enzyme that is not the active site; this changes the enzyme's tertiary structure, including the shape of the active site, so the substrate can no longer fit	• Potassium cyanide binds to haem, which is part of cytochrome oxidase, an enzyme of respiration (non-reversible)

* Inhibitors can be reversed if the inhibitor does not bind permanently to the enzyme.

✓Quick check 3, 4, 5

QUICK CHECK QUESTIONS

1. Explain the effect of temperature on enzyme activity.
2. Explain why enzymes often function efficiently only over a narrow pH range.
3. Name: (i) a metal ion that acts as a cofactor; (ii) an organic compound that acts as cofactor; (iii) a coenzyme.
4. Explain how potassium cyanide acts as a metabolic poison.
5. Explain the effect of a named drug that acts as an enzyme inhibitor.

Experiments with enzymes

Key words

- rate of reaction
- collisions
- substrate
- catalase
- amylase

Examiner tip

When describing enzyme-catalysed reactions write about the *collisions* between substrate molecules and the enzyme's active site.

Hint

Remember that in both examples described here results are taken at intervals of time from the *same reaction mixture*.

When an enzyme and its **substrate** are mixed together, a reaction begins. Substrate molecules **collide** with the enzyme and bind to its active site. Product molecules are formed. As the reaction proceeds, the number of substrate molecules decreases and the number of product molecules increases. The number of enzyme molecules stays constant. The speed, or **rate**, of the reaction can be followed by measuring either

- increasing quantities of product, or
- decreasing quantities of substrate.

In this spread we will look at an example of each.

Increasing product

The figure below shows some apparatus that may be used to measure the activity of the enzyme **catalase** in breaking down hydrogen peroxide:

$$2H_2O_2 \rightarrow O_2 + 2H_2O$$

The rate of the reaction is determined by measuring how much oxygen is collected at intervals of time and using this formula:

$$\text{rate of reaction} = \frac{\text{volume of oxygen produced}}{\text{time}}$$

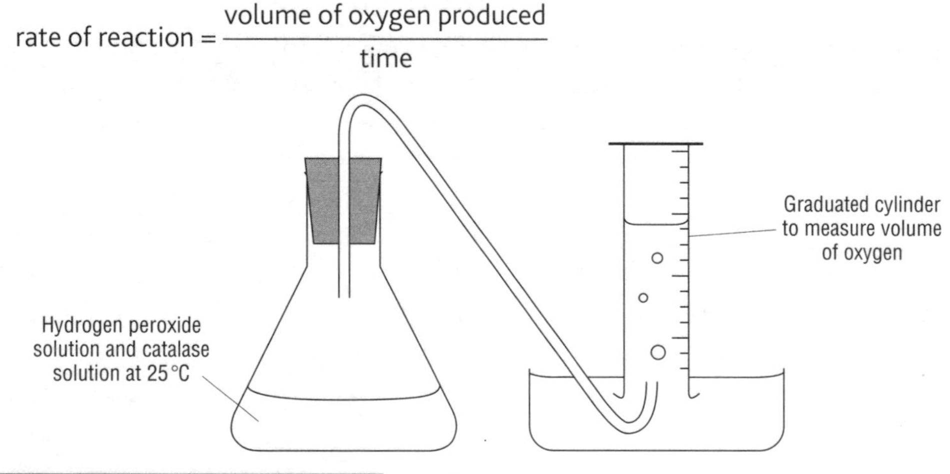

time/s	volume of oxygen/cm^3
30	3.0
60	5.0
90	6.0
120	6.5
150	6.9
180	7.0
210	7.0

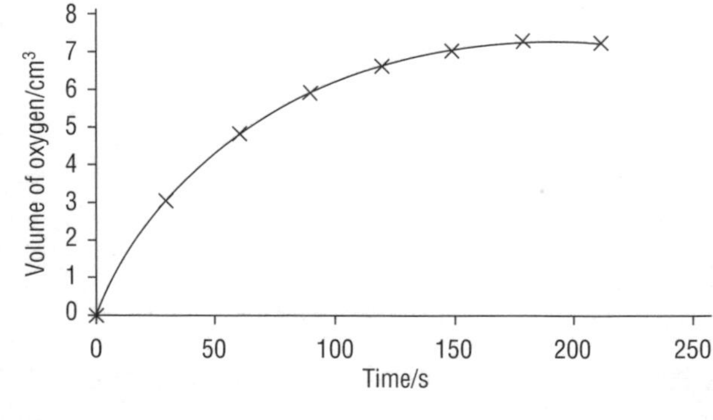

The figure above shows these results plotted on a graph.

As the reaction proceeds, less oxygen is produced as there is less substrate available. The rate of reaction is quickest at the beginning when there is a high concentration of substrate. Later, substrate concentration becomes a limiting factor so the reaction slows down. Eventually all the substrate is used up so the reaction stops.

Decreasing substrate

Plants and animals break down starch to maltose using the enzyme **amylase**.

$$\textbf{starch + water} \xrightarrow{\text{amylase}} \textbf{many molecules of maltose}$$

The apparatus is used to follow the breakdown of starch. Samples are taken from the reaction mixture and tested with iodine solution. At the beginning, there is plenty of starch in the reaction mixture and the colour with iodine is dark blue. Later, most of the starch has been hydrolysed and the colour is light blue. When all the starch is broken down the colour is yellow. The colorimeter measures the absorbance of samples. A dark colour with iodine solution (indicating much starch present) gives a high absorbance reading.

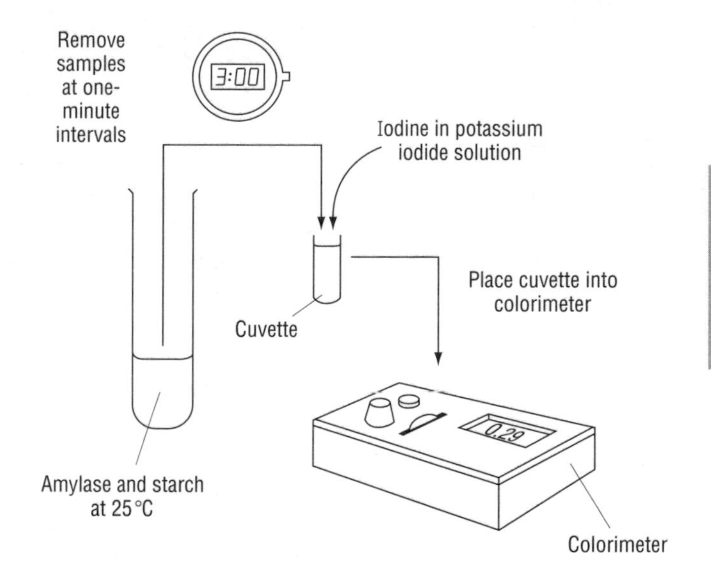

time/min	absorbance/arbitrary units
0	2.00
1	1.01
2	0.55
3	0.29
4	0.23
5	0.19
6	0.16
7	0.11
8	0.07
9	0.05
10	0.00

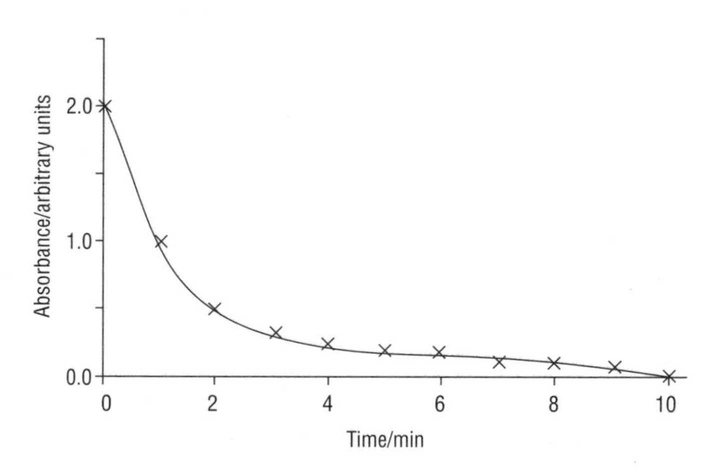

The graph shows absorbance readings decreasing rapidly at the start and more slowly later on. At the start there are many molecules of starch and all the active sites of the enzyme are filled. Later as the concentration of starch falls, there are many active sites unfilled and the rate of reaction falls. Known concentrations of starch are tested in the colorimeter. The absorbance readings are plotted on a graph, which is used to convert the results from the table into actual concentrations of starch. The concentrations of starch are then plotted against time on a graph; gradients of the curve are calculated to give the rate of hydrolysis of starch at different times.

Quick check 1, 2, 3

QUICK CHECK QUESTIONS

1. What happens to the substrate concentration during the course of an enzyme-catalysed reaction?
2. Explain why the reaction catalysed by amylase occurs quickly at the start and then slows down.
3. What do you think would happen to the reactions catalysed by catalase and amylase if they were carried out: (a) at 35 °C and not 25 °C; (b) with higher concentrations of enzyme?

Balanced and unbalanced diets

Key words

- balanced diet
- malnutrition
- obesity
- prevalence
- risk factor
- atherosclerosis

We need to eat a balanced diet that is related to our needs, as listed in the table below.

Requirement	Comments
Sufficient energy for our needs	Provided by the macronutrients (carbohydrates, fats and proteins); energy intake by the body should equal energy expenditure
Essential amino acids	Our metabolism cannot make these amino acids from anything else – they have to be in the protein we eat
Essential fatty acids (linolenic acid and linoleic acid)	Our metabolism cannot make these either – they have to be in foods containing fat or oil
Micronutrients – vitamins and minerals	Required for a wide variety of functions: many B-group vitamins are used to make coenzymes; some minerals (e.g. Zn and Cu) are cofactors (see page 57)
Water to replace that lost in urine, sweat, breath, faeces	Water has many functions in the body (e.g. as a solvent, a reactant and a coolant); cytoplasm is about 70% water
Fibre	Prevents constipation; helps reduce risk of heart disease and bowel cancer

If you read through pages 38 to 57 you will find out how the components of our diet are used in our metabolism. For example:

- fats are used to make phospholipids and cholesterol for cell membranes
- carbohydrate is stored as glycogen in liver and muscles
- proteins in the diet are hydrolysed to give amino acids, which are then used to make our own proteins, such as haemoglobin, collagen and all our enzymes.

Malnutrition

Malnutrition means eating much less – or much more – than needed.

✓*Quick check 1*

- People who are starving do not have sufficient energy or nutrients, and often show symptoms of protein–energy malnutrition. Deficiencies of specific nutrients impair health; for example, a lack of vitamin D leads to rickets.
- Eating more than is needed can lead to obesity, which is associated with many risks to health such as cancer, type 2 diabetes and coronary heart disease (CHD). A person who is very overweight is obese.

Body mass index (BMI) is a way of determining whether a person is overweight. This is calculated as:

✓*Quick check 2*

BMI = body mass in kg / (height in m)^2

BMI	Category
Less than 18.5	Underweight
18.5–24.9	Acceptable
25–29.9	Overweight
30–34.9	Obese (class 1)
35–35.9	Obese (class 2)
Over 40	Severely obese (class 3)

Examiner tip

Be careful here about *mass* and *weight*. People talk about units of kilograms (or stones and pounds) as *weight*, when in scientific terms they should say *mass*.

The **prevalence** of obesity is increasing in affluent countries as people eat far more food than they need and take less exercise.

Diet and coronary heart disease

Many factors, such as diet, lack of exercise, obesity and heredity, contribute to the development of CHD. The prevalence of CHD is high in some countries, for example Scotland, where people consume large quantities of animal fats rich in saturated fatty acids. Countries where people have low consumption of animal fat, such as Japan, have a very low prevalence of CHD. People with high blood pressure (hypertension) are at high risk of developing CHD. Hypertension is related to a high intake of salt that lowers the water potential of blood. Water is retained in the blood, so increasing blood volume, which may lead to hypertension. This puts a strain on arteries in the brain, leading to a stroke. The extra work that is done by the heart may put it under strain and contribute to developing heart disease. Eating highly refined foods with high sugar content and not enough fibre are also **risk factors** for CHD.

Examiner tip

In this case, prevalence means the number of people in a population who are obese. In England in 2007, 17% of men and 21% of women were estimated as being obese (BMI > 30).

Hint

The structure of saturated and unsaturated fatty acids is shown on page 46.

Blood cholesterol

Lipoproteins are small particles made in the liver to transport cholesterol. Each lipoprotein is coated with protein to make it water-soluble, and contains a core of cholesterol and other lipids. Cells with appropriate receptors can absorb lipoproteins by endocytosis. There are two groups of lipoprotein:

- low-density lipoprotein (LDL)
- high-density lipoprotein (HDL).

LDLs deliver cholesterol to tissues; HDLs remove cholesterol from tissues and return it to the liver.

✓Quick check 3

If the endothelial lining of an artery is damaged, LDLs tend to deposit cholesterol, which accumulates along with fatty acids, calcium salts and fibrous tissue. This fatty material forms **plaques** (**atherosclerosis**) inside the wall of the arteries. Atherosclerosis is the progressive build-up of plaque. Plaque enlarges the wall so that there is less space for blood to flow. It also roughens the lining of arteries, so increasing the chances of blood clots forming.

Examiner tip

Plaques develop *in* artery walls, not *on* the walls. Do not refer to arteries as blood vessels – that is too vague.

Oxygenated blood flows from the aorta into the coronary arteries, which supply the muscle in the atria and ventricles of the heart. If these arteries are narrowed with plaques, the heart muscle may be deprived of oxygen and become fatigued. This means that people with this condition find even mild exercise difficult and have chest pains that disappear when they stop exercising. This form of CHD is angina. If a blood clot forms in a coronary artery it may reduce blood flow so much that heart muscle dies, thus causing a heart attack that may be fatal.

The risk of heart disease is increased if there is:

- a concentration of cholesterol in the blood greater than 5 mmol dm^{-3}
- a high concentration of LDLs (> 3 mmol dm^{-3})
- a low concentration of HDLs (< 1 mmol dm^{-3})
- a low ratio of HDLs to LDLs. The ideal ratio is 4 : 1.

QUICK CHECK QUESTIONS

1 Explain the term *malnutrition*.
2 Describe how to determine whether someone is obese.
3 Explain why cholesterol is transported in the form of lipoprotein.

Food production and preservation

Key words

- autotroph
- heterotroph
- selective breeding
- artificial selection
- fertilisers
- mycoprotein
- food spoilage
- pasteurisation
- irradiation
- sterilisation

Food chains

Producers form the first trophic level of food chains. Producers are **autotrophs** – they use an external energy source and simple inorganic molecules to make complex organic molecules. All the other organisms in the food chain are **heterotrophs** – organisms that take in complex organic molecules (such as carbohydrates, proteins and lipids) to act as a source of energy and to use in their metabolism.

Your diet depends on plants. Many foods, such as bread, are processed plant material. Vegetables and fruits are parts of plants. Animals and animal products, such as milk, cheese and eggs, come from the second trophic level of food chains. Some foods, mostly fish and fish products, come from the third or even fourth trophic level (food chains in the sea tend to be longer than those on land because the producers are tiny).

Selective breeding

Domesticated animals have been subjected to **artificial selection**. Breeders choose the feature or features they wish to improve. Individuals that show the desired features are selected and bred together. The offspring that show an improvement are selected to breed the next generation. Changes occur over many generations, giving varieties very different in appearance from the original stock, and with much improved productivity.

A plant breeder may breed a high-yielding variety susceptible to a fungal disease with a low-yielding variety resistant to the disease by transferring pollen. Seed is collected and sown in the next growing season. The offspring are tested, and if they are disease-resistant they are crossed with the high-yielding variety to maximise the contribution of that variety. This may continue for up to 10 years.

Some achievements of selective breeding of crops and livestock are shown on the left.

Crop plants

- Increased yield of grain, roots, tubers (potatoes), etc.
- Resistance to pests, e.g. insect pests
- Resistance to fungi, bacteria and viruses
- Better quality, e.g. nutritional value and flavour

Livestock

- Increased yield of meat, milk, eggs, etc.
- Fast-growing breeds
- Resistance to disease (e.g. blue tongue disease)
- Better quality, e.g. lean (low-fat) meat

Quick check 1

Environment matters too

Only so much can be achieved by breeding for improvement. Productivity has increased because farmers have improved the environments of their livestock and crop plants.

When crops are harvested, only part of the plant is left to decompose and return nutrients to the ground. Yields decrease with time if soil fertility is not restored with fertilisers. Organic fertilisers are the wastes of animals and composted plants; artificial fertilisers contain the chemicals crop plants need as nutrients in the correct proportions:

- the element nitrogen, as nitrate or ammonium ions, to make amino acids
- magnesium ions to make chlorophyll
- potassium ions as enzyme **cofactors** and for guard cells to open stomata
- phosphate ions to make DNA, RNA, and coenzymes such as NADP.

Numerous pests and diseases consume crop plants, and weeds compete with crops for light, water and nutrients. Farmers can use various **pesticides**.

- Herbicides are applied before the crop germinates to kill weeds that compete with freshly germinated crop plants.
- Fungicides are applied if weather conditions make it likely that the crop will be infected by fungi.
- Insecticides are applied when insect pests reach a level that will cause economic loss.

Organic farmers do not use pesticides; instead they use methods such as crop rotation and natural predators of pests (biological control).

Antibiotics are used to treat livestock that become ill. They can also be added to animal feed to reduce the activity of gut bacteria so more food is available to the animals – a practice now banned in the EU, although it continues in the USA.

Foods from microorganisms

Some bacteria and fungi were 'domesticated' by humans thousands of years ago. These microorganisms make foods using biological processes that we now understand and can modify.

- Bacteria are used to make yoghurt.
- Bacteria and fungi are used in cheese-making.
- Yeasts (single-celled fungi) ferment sugar to make alcohol in brewing and wine-making.
- Yeasts ferment to produce carbon dioxide that makes dough rise for bread-making.

Mycoprotein is a new food that uses microorganisms in its production. *Fusarium venenatum* strain PTA-2684, a soil mould fungus, was discovered in a field in Buckinghamshire in 1967. Since 1985 it has been used on an industrial scale to make a meat substitute marketed as Quorn™. Microorganisms are also used to make food additives, such as vitamins and the flavour enhancer monosodium glutamate.

The advantages and disadvantages of using microorganisms in foods are listed on the right.

✓ Quick check 2

Food spoilage

Our foods can be contaminated by microorganisms that feed as decomposers. The damage done by food-spoilage organisms ranges from mould growing on bread to large silos of grain being affected by the fungus *Aspergillus*, which produces **carcinogenic** toxins called aflatoxins. Food-spoilage organisms make food appear 'off', taste awful and smell bad, and may be harmful to our health. In order to grow, they need organic material (our food), water, a suitable temperature, oxygen (usually) and a suitable pH. Food preservation techniques remove one or several of these conditions. Many methods to preserve foods have been used for thousands of years; several new methods have been developed in the past 100 years or so. The table shows some of these methods and the biological principles behind them.

✓ Quick check 3

Microorganisms

Advantages

- Microorganisms grow quickly, giving high yields and fast production
- Factories use less land than traditional agriculture, and can be set up anywhere
- Can use waste material as substrate
- Selection and genetic engineering are easier than with animals and plants
- No ethical issues (as with keeping livestock)
- Low-fat or no-fat foods

Disadvantages

- Microorganisms are subject to infections (e.g. by viruses), with loss of production
- Production vessels (fermenters) can be contaminated by competitors (e.g. bacteria), with loss of production
- Customer resistance to new foods
- Fungi, yeasts and bacteria must have substrates, such as sugar or starch, produced by plants
- Purification before entering the human food chain may be expensive, for example removal of nucleic acids

Food preservation method	Example	Biological principle
Salting	Salted cod	Salt removes water from organisms by osmosis; using sugar for preservation uses the same principle
Pickling	Sauerkraut (pickled cabbage)	Ethanoic acid (vinegar) gives a low pH (<4.0) so that enzymes in spoilage organisms are denatured
Heat treatment	Milk, wine	• Heated to 71.7 °C for 15 seconds (pasteurisation), so killing potential pathogens but not all bacteria. Flavour is preserved • Heated to at least 135 °C for at least one second so killing all bacteria (UHT). This changes the flavour
Freezing	Meat	Water is frozen, so is not available to organisms. Enzymes are inactive
Irradiation	Fruit, prawns	X-rays/gamma-rays kill bacteria and moulds by breaking bonds in proteins and DNA

QUICK CHECK QUESTIONS

1 Describe the principle of selective breeding.

2 Name five foods made by microorganisms.

3 Explain the term *food spoilage* and describe how food may be prevented from going 'off'.

Health and disease

Module 2

Key words

- pathogen
- parasite
- disease
- transmission
- opportunistic infections

Examiner tip

Make sure you know about the methods of transmission and how this knowledge helps us to develop methods to control disease.

The World Health Organization (WHO) defines **health** as a state of complete physical, mental and social wellbeing, which is more than just the absence of disease. **Disease** may be defined as an absence of health, but we tend to use the word to refer to specific states of bad health that give certain symptoms that we experience. We report these symptoms to doctors, who look for certain clinical signs to decide which disease we have.

We are afflicted by numerous diseases that can be catalogued into different groups: inherited and non-inherited; chronic and acute; infectious and non-infectious, and so on. This spread is concerned with infectious diseases caused by **parasites** with which we do not live in harmony. Our bodies are hosts to many organisms – countless numbers of bacteria live on our skin, inside our mouth and inside our guts. Most of these are not harmful, although those that cause teeth decay certainly are. We may also be the host for larger organisms, such as lice, ticks and fleas. Parasites are organisms that live in or on a host and obtain their nourishment from their host.

Some of the bacteria described above provide us with some useful services, such as producing substances we need (e.g. vitamin K) and successfully competing with harmful bacteria so they do not infect us.

Pathogens are parasites that invade the body, multiply in tissues or inside cells and cause disease. Disease **transmission** is the transfer of a pathogen from infected to uninfected people.

- Malaria and tuberculosis (TB) are caused by pathogens that invade our cells and then spread through the tissues.
- Human immunodeficiency virus (HIV) infection can lie dormant in T lymphocytes (see page 66) in the body for a long time, but eventually weakens the immune system so that people become susceptible to **opportunistic infections**, such as pneumonia, and certain cancers. The collection of these diseases is known as AIDS (acquired immune deficiency syndrome).

Causative organisms and means of transmission

Disease	Causative organism (pathogen)	Main methods of transmission
Malaria	Protoctist: several species of *Plasmodium*	Insect vector: female *Anopheles* mosquito
TB	Bacterium: *Mycobacterium tuberculosis*, *Mycobacterium bovis*	Airborne droplets of water *M. bovis* in milk and meat from infected cattle
HIV/AIDS	Virus: human immunodeficiency virus	• During unprotected sexual intercourse • Infected blood and blood products • Sharing or re-using hypodermic needles • Across placenta from mother to fetus • Blood-to-blood contact from mother to baby at birth

✓Quick check 1

Global impact of infectious diseases

Disease	Global distribution
Malaria	Widely distributed throughout the tropics and sub-tropics
TB	Worldwide – throughout developing world, Russia and Central Asia; among homeless and poor in inner cities in developed world, especially among immigrants from developing countries
HIV/AIDS	Worldwide: highest prevalence in sub-Saharan Africa and South-east Asia

✓Quick check 2

Measures can be taken to control and prevent the spread of disease. This is done by breaking the transmission from infected to uninfected people.

Disease	Global impact	Control measures
Malaria	• 40% of world population live in malarial areas • More than 500 million people become ill with malaria each year • More than 1 million people die of malaria each year – mostly infants and children in Africa • Mosquitoes are resistant to insecticides • *Plasmodium* is resistant to drugs (such as chloroquine) that are used to kill it	• Prevent mosquitoes biting at night by using sleeping nets (most effective when nets are soaked in insecticide) • Use drugs that prevent *Plasmodium* spreading through the body • Reduce mosquito populations by spraying insecticide or putting fish in ponds, streams, irrigation ditches
TB	• One-third of human population carry the bacterium although in many it is inactive • 8.8 million new cases and 1.5 million deaths each year • Disease is more likely to spread in poverty-stricken and overcrowded conditions • Spread of MDR-TB (multiple drug resistant) and XDR-TB (extensively drug resistant). XDR-TB is resistant to 2 of the 3 'first-line' drugs and 2 of the 'second-line' drugs used when the 'first-line' drugs have failed	• Cured by long course of antibiotics, but some people do not finish the course • Directly observed treatment, short course (DOTS) is a WHO strategy against TB (volunteer supervises patient to make sure antibiotics are taken daily and the course is completed) • BCG vaccine against TB is not very effective – routine use discontinued in UK in 2005
HIV/AIDS	• 39.5 million people living with HIV/AIDS (25 million in sub-Saharan Africa); 4.9 million new HIV infections in 2006 with 2.9 million deaths from AIDS • A disease both of poverty and affluence – all social groups affected • TB is an opportunistic disease associated with HIV	• Using condoms or femidoms during sexual intercourse • Health education about safer sex • Contact-tracing to find people likely to be infected • Blood donations screened for HIV (although false negatives may be a problem) • Blood donations and blood products heat-treated to kill viruses • Needle exchange schemes

✓Quick check 3

QUICK CHECK QUESTIONS

1 Name the organisms that cause malaria and tuberculosis (TB), and explain how these two diseases are transmitted from infected to uninfected people.

2 Describe the global distribution of malaria, TB and HIV/AIDS.

3 Describe how the spread of malaria, TB and HIV/AIDS may be controlled.

The immune system

Key words

- phagocytes
- macrophages
- immune response
- lymphocytes (B and T)
- antibody
- antigen
- memory cell
- active immunity
- passive immunity

Module 2

Quick check 1

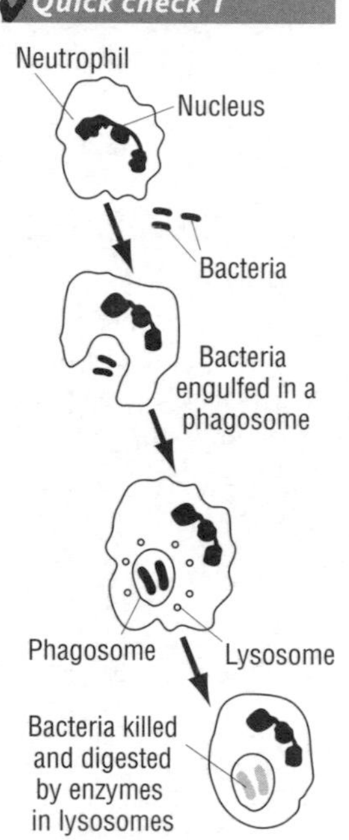

A neutrophil carrying out phagocytosis

The immune system

Before and shortly after we are born, our developing immune system gains the ability to distinguish between 'self' (our own cells) and 'non-self' (anything foreign including pathogens and the toxins they produce). The immune system defends the body against infectious diseases. The **primary defences** prevent **pathogens** from entering tissues. They are:

- the epidermis, layers of dead skin cells containing the fibrous protein **keratin**
- mucus secreted by the epithelial lining of the airways, digestive system and reproductive system to trap bacteria and other particles; the airways also contain ciliated cells to move the mucus to the mouth, where it is swallowed
- hydrochloric acid secreted by the stomach lining to kill most organisms we ingest (mucus protects the stomach lining from the acid).

Pathogens that pass these barriers can feed, grow and divide. They may spread beyond the original site of infection through the blood or lymphatic system. The next line of defence is formed by the five groups of cells shown in the table.

Cell type	Distribution	Function
Neutrophil	Blood and tissues	Phagocytosis
Macrophage	Tissues (e.g. lungs)	Phagocytosis
B lymphocyte (B cell)	Blood and lymph nodes	Production of antibodies
T helper lymphocyte (T_h cell)	Blood and lymph nodes	Stimulate B lymphocytes to divide and to produce antibodies; stimulate phagocytosis
T killer lymphocyte (T_k cell)	Blood and lymph nodes	Destroy cells infected with viruses

The immune response

Phagocytes, such as **neutrophils** and **macrophages**, are not very successful alone. Alongside **lymphocytes** in an immune response they are much more effective. The body contains many millions of B and T lymphocytes, especially in lymph nodes and in the spleen, bone marrow and thymus. These cells have glycoproteins on their surfaces that each recognise a different antigen (usually a protein or glycoprotein) on the surface of a specific pathogen. When a pathogen enters the body, its antigens bind to the B and T lymphocytes that have glycoproteins of complementary shape, stimulating an immune response. These B and T lymphocytes divide by mitosis to form many identical cells. Some become effector cells; others become **memory cells**. B lymphocytes that become effector cells (**plasma cells**) make and release lots of **antibodies**.

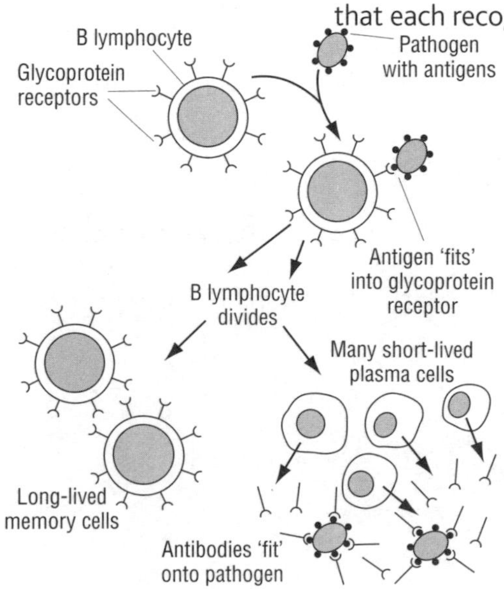

The events that occur during an immune response. Note that antibody molecules are very much smaller than the cells that secrete them

The immune response is coordinated by local hormones known as cytokines. The main cell type involved is the T helper lymphocyte, which releases cytokines to stimulate B lymphocytes to divide. This is an example of cell signalling (see page 10). The events shown in the diagram occur during the first (primary) immune response to an invasion by a **pathogen**. If a pathogen with the same antigens invades again, the secondary response is much faster, because this time there are memory cells that can divide and become effector cells quickly. The pathogen is destroyed before it can cause disease.

Immunity	Active	Passive
Natural	*Advantage:* long-term immunity *Disadvantages:* immune response takes time; symptoms develop; disease may be fatal *Example:* catching measles provokes an immune response	*Advantage:* immediate protection against diseases to which the mother has active immunity *Disadvantages:* short-term; no memory cells produced *Example:* antibodies passed from mother to child across placenta and in colostrum (breast milk)
Artificial	*Advantages:* long-term immunity; no need to suffer from the disease *Disadvantage:* immune response takes time *Example:* vaccination against measles or tetanus	*Advantage:* immediate protection against specific disease *Disadvantages:* Short-term; no memory cells produced *Example:* antibodies against tetanus toxin collected from blood donations and injected

Module 2

People are immune to a disease when they can mount a fast, effective defence against the pathogen so no symptoms develop. Immunity arises in one of two ways.

- **Active immunity**: the immune system develops antibodies directed against the specific pathogen. It is usually long-term, as memory cells are produced.
- **Passive immunity**: antibodies are transferred from another source. The immune system does not produce its own antibodies or memory cells, so the immunity lasts only weeks, or months at most, as the foreign antibodies are destroyed.

Both types of immunity can be natural or artificial. The table shows how these four forms of immunity are gained, and their advantages and disadvantages.

✓Quick check 2

Host cell

(1) Bacteria secrete toxin

(2) Antibodies combine with (neutralise) toxin

(3) Host cells protected from toxin

Antibodies link bacteria in clumps that can be found and engulfed by phagocytes (agglutination)

How antibodies act by neutralisation and agglutination

Antibodies

During an immune response B lymphocytes develop into plasma cells, which make glycoproteins called antibodies. Antibodies consist of four polypeptides and have two or more antigen-binding sites. These variable regions are different shapes in different antibodies to complement the particular antigen for which an antibody is specific. The constant region is the same shape in all antibodies of the type shown in the diagram.

Antibodies work by neutralisation and agglutination, as shown in the diagram. Neutralisation means that antibodies combine with viruses, to stop them entering cells, or with toxins released by some pathogens, to make them harmless. Bacteria are agglutinated into clumps that phagocytes can recognise.

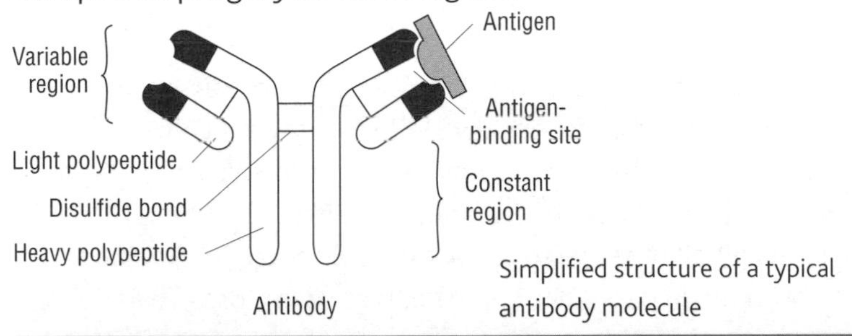

Simplified structure of a typical antibody molecule

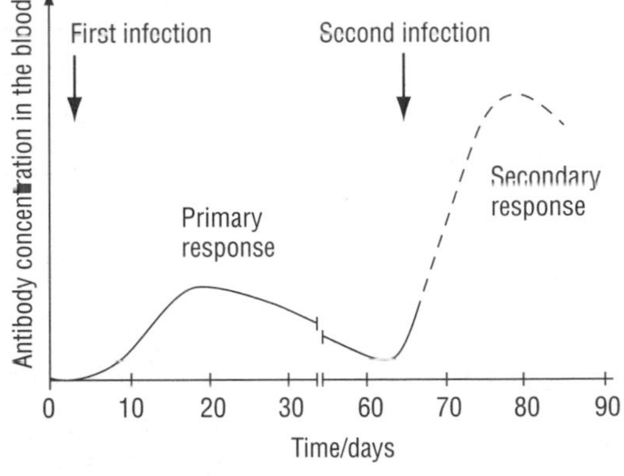

The changes in antibody concentration during primary and secondary responses to an infection

QUICK CHECK QUESTIONS

✓Quick check 3

1 Describe the mode of action of B lymphocytes and T lymphocytes.

2 Describe the structure of an antibody molecule and explain how it is specific to an antigen.

3 Draw a graph to show what you predict will happen to the antibody concentration in an infant's blood following the first vaccination for polio at 2 months, and following boosters at 3 and 4 months.

Vaccines and other medicines

Key words

- vaccine
- vaccination
- herd immunity
- ring vaccination
- antibiotic

We saw on page 67 that active immunity may be gained artificially by vaccination. A **vaccine** is a preparation of an antigen or many antigens for a specific infectious disease, and is injected or given by mouth.

Some vaccines contain live, but weakened, forms of the pathogen.

- The vaccine for measles, mumps and rubella (MMR) contains attenuated (weakened) viruses that can infect cells, but are harmless.
- The BCG vaccine against TB is an attenuated form of *Mycobacterium bovis*, the pathogen that causes the disease in cattle and can be caught by humans.

✓Quick check 1

A vaccine stimulates a primary immune response (see primary and secondary responses in the graph on page 67), so it takes time to gain immunity. The response is good if the vaccine is a 'live' vaccine containing living organisms. This is possibly because the weakened form of the pathogen remains in the body and may multiply, so there is a longer exposure to the immune system, giving a better response. To improve the response to vaccines, boosters are given to stimulate secondary responses and the development of larger clones of lymphocyte memory cells specialised to act against the pathogen concerned.

Vaccination success stories

Smallpox was the first disease to be eradicated from the world. It is likely that polio will be the next. Since 1988, WHO has coordinated mass immunisation programmes against polio, making this the largest public health initiative to date.

This form of immunisation is known as **herd immunity**. It is effective because if anyone is infected by the virus, develops symptoms and becomes infectious, there is nowhere for the virus to go, as nearly everyone is immune. With time, polio has become restricted to India, Pakistan and Nigeria, and vaccination programmes are concentrated on those areas. Occasionally, cases of polio occur elsewhere – possibly because someone has travelled from an area where the disease still exists, or sometimes because the live (attenuated) vaccine regains virulence. In these cases, **ring vaccination** is used: everyone in the surrounding area is vaccinated to prevent transmission.

✓Quick check 2

Influenza

Influenza is a viral disease. The virus infects the lining of the airways and is associated with fever, sore throat, headache, muscle pains and weakness. In severe cases it can lead to pneumonia and may be fatal. There have been some very serious outbreaks of influenza, and it is a disease that WHO and national health authorities monitor very carefully. Thorough surveillance is carried out all the time to check for newly emerging strains of the virus. New strains often originate in East Asia, where people live in close proximity to domesticated animals such as pigs and chickens, which also harbour the virus.

The H5N1 strain of the virus that causes influenza in birds is an indication of what may happen when the influenza virus mutates into a new strain. As yet, H5N1 is not transmitted easily between people.

The virus that causes human influenza may 'cross-breed' with viruses that cause similar diseases in animals, or a strain that is pathogenic in animals may cross the species barrier and infect us. 'Cross-breeding' occurs when viruses of two strains infect the same cell. As each new virus particle buds from the host cell it takes genes from both strains. Each year, WHO advises countries to vaccinate people at risk with specific vaccines against the three strains that are predicted to infect people that year. In the UK, all people over the age of 65 are offered influenza vaccination, as are young people with conditions such as asthma, and people in high-risk categories such as medical staff and carers. Herd immunity may become necessary if a new, dangerous strain emerges. However, it is unlikely that sufficient stocks of the required vaccine would be ready in time to vaccinate most of the population.

✓Quick check 3

New medicines

Pharmaceutical companies are always developing new medicines. There are several reasons for this.

- Pathogens have become resistant to existing drugs, such as many **antibiotics**.
- New diseases have emerged over recent years, and there will be more in the future.
- Vaccines are needed for many diseases, and existing vaccines can be improved.

Various fungi and bacteria, such as *Streptomyces*, are the source of most antibiotics, although many are modified chemically before they are formulated and sold. Pharmaceutical companies invest in searches for potential new medicines. They search soil samples from all over the world in the hope of finding microorganisms that produce antimicrobial substances that could become antibiotics of the future. They screen compounds extracted from plants – especially those used in traditional medicines. Currently much interest is being shown in the plants used in traditional Chinese medicine. Plants have complex metabolisms that produce a very wide range of chemicals, some of which may have potential as a medicine. The table shows four such medicines.

Medicine	Plant source	Use	Biological action
Taxol®	*Taxus* spp., yew tree	Anticancer agent	Inhibits mitosis by interacting with microtubules
Vinblastine Vincristine	*Catharanthus roseus*, Madagascan periwinkle	Anticancer agent	Inhibit mitosis by interacting with microtubules
Artemisinin	*Artemisia annua*, sweet wormwood	Antimalarial drug	Attacks various stages of *Plasmodium*

These four drugs were originally found in plants, although they are now synthesised chemically. Without the plants, we would not have discovered them. It is therefore important that plant species do not become extinct (see page 85).

✓Quick check 4

QUICK CHECK QUESTIONS

1. Explain what is meant by a vaccine.
2. Outline the ways in which vaccination is used to control the spread of disease.
3. Outline the ways in which health authorities control influenza.
4. In the context of the information in this spread, explain why it is important to preserve species from extinction.

Smoking and disease

Key words

- chronic bronchitis
- emphysema
- carcinogen
- carboxyhaemoglobin
- epidemiology

Examiner tip

Make sure you recognise that cardiovascular system = heart and blood vessels.

The World Health Organization (WHO) considers smoking to be an epidemic. This is because smoking is the cause of numerous diseases, and a contributory factor to many more. The major effects are on the **gas exchange system** (trachea, bronchi and lungs) and the **cardiovascular system** (heart and blood vessels).

When dried tobacco leaves are burnt, they give off a large number of substances that are inhaled into the lungs. Some remain in the lungs; others are absorbed into the blood. This table shows some of these substances and their effects on the body.

Quick check 1

Substance	Effects on the body
Tar	• Accumulates in the airways (especially the bronchi) • Destroys cilia • Stimulates goblet cells to secrete more mucus • Causes chronic bronchitis and emphysema
Carcinogens	• Cause mutations to occur in bronchial epithelial cells, leading to formation of tumours (lung cancer)
Carbon monoxide	• Absorbed into blood • Combines with haemoglobin to form carboxyhaemoglobin • Reduces oxygen-carrying capacity of the blood and starves heart muscle of oxygen
Nicotine	• Absorbed into the blood • Increases heart rate and blood pressure, causing damage to artery walls • Stimulates decrease in blood flow to extremities • Increases chances of blood clots forming

Chronic obstructive pulmonary disease and lung cancer

Chronic obstructive pulmonary disease (COPD) is a disease associated with smoking, and includes the conditions **chronic bronchitis** and **emphysema**, which are described in the table below.

But for smoking, lung cancer would be a very rare disease. Cancer-causing agents (**carcinogens**) in tobacco smoke cause mutations in the epithelial cells lining the bronchi that may eventually lead to the growth of a tumour.

Quick check 2, 3

Disease	Changes in the lungs	Symptoms
Chronic bronchitis	Bronchi become obstructed and narrow because: • lining is inflamed • smooth muscle layer thickens • goblet cells and mucous glands produce much mucus	Shortness of breath Wheezing Persistent cough
Emphysema	• Alveoli become overstretched, lose elasticity and burst • Fewer elastic fibres • Large gaps in the lungs, giving smaller surface area for gaseous exchange	Shortness of breath Difficulty in breathing out In severe cases people need to breathe oxygen through a mask
Lung cancer	• Bronchi blocked by cancerous growths	Coughing up blood; persistent cough; weight loss

Epidemiological evidence

Epidemiology is the study of patterns of disease. The link between cigarette smoking and lung cancer was first suggested in the 1950s by epidemiologists who collected data from patients with the disease. They found that almost all lung cancer patients were smokers. Many studies since then have shown that smoking is the major cause of lung cancer. Very few non-smokers develop the disease. Many smokers develop it and die from it. The table below summarises some of this epidemiological evidence.

Observations	Explanation
Lung cancer was a rare disease before the twentieth century	Cigarettes were first made at the end of the nineteenth century; smoking became common early in the twentieth century
Cases of lung cancer increased from the 1930s onwards	Smoking became common during the First World War; it takes 20–30 years for symptoms to develop
More men than women suffer from lung cancer	For most of the twentieth century, more men than women smoked cigarettes
Most people who develop lung cancer are smokers	Tar from cigarette smoke contains carcinogens (other causes of lung cancer are very rare)
Death rates from lung cancer are highest among people who smoke more than 25 cigarettes a day	People who smoke many cigarettes in a day expose their lungs to more carcinogens, increasing the chance of cancer

Experimental evidence

There are two main lines of experimental evidence for the link between smoking and lung cancer:

- dogs that were exposed to cigarette smoke in the same way as humans developed cancerous growths in their lungs
- when substances extracted from tar in cigarette smoke were painted on the skin of mice, tumours started to develop.

These experiments show that cigarette smoke contains carcinogens that cause genes to mutate so that cells start to divide uncontrollably to give cancerous growths or tumours.

Examiner tip

You are expected to *evaluate* the evidence concerning lung cancer. You should look to see whether or not the evidence shows causality (cause and effect). Epidemiological evidence does not – it shows correlations that may be coincidental, or related to a third factor.

✓ *Quick check 4*

QUICK CHECK QUESTIONS

1 Distinguish between the gas exchange and cardiovascular systems.

2 Describe the effects of cigarette smoke on the gas exchange system.

3 Describe the effects of cigarette smoke on the cardiovascular system.

4 Describe the epidemiological and experimental evidence that confirms the link between cigarette smoking and lung cancer.

Biodiversity

Key words
- biodiversity
- ecosystem
- habitat
- species

Imagine you are out in space, looking at the Earth. You can see our blue planet that supports life – the biosphere. Now focus on areas of blue, yellow, green and white. These are areas of ocean, desert, grassland and forest, and ice and snow. These are the major ecological areas that are characterised by their physical environment or their dominant vegetation, for example deserts and grassland. Within these ecological areas are **ecosystems**, which consist of communities of plants, animals and microorganisms that interact with each other and their physical environment. Ecosystems can be as large as the Mississippi delta or as small as a pond. The **biodiversity** of an area is a measure of the:

- different ecosystems
- number of species
- number of individuals of each species
- genetic variation within each species present in an area.

✓*Quick check 1*

The **habitat** of an organism is the place where it lives. It may be the name of that place (e.g. Aldabra, the habitat of the Aldabran Giant Tortoise, *Geochelone gigantea*), or it may be a description of the dominant vegetation (such as coniferous forest, the habitat of the crossbill, *Loxia curvirostra*, a species of bird found in northern latitudes), or it may be a type of environment (e.g. a pond is the habitat for the water hog louse, *Asellus aquaticus*). A full description of a habitat includes the physical and biological factors that characterise that environment.

✓*Quick check 2*

It is difficult to give a definition of the term **species** that satisfies all biologists. There are two groups of criteria that biologists use to define the term.

1. A group of more-or-less similar organisms that are:
   - capable of interbreeding
   - capable of producing fertile offspring
   - reproductively isolated from other groups.
2. A group of organisms that show close similarity in a number of characteristics:
   - morphological (outward appearance)
   - physiological (how their body functions)
   - embryonic (how their body grows and develops)
   - ecological (where they live; how they interact with their environment and other organisms)
   - behavioural (how they feed, move and interact with members of the same group).

A 'biospecies' fulfils the first set of criteria. Using this definition, it is possible to tell if two organisms belong to the same species only if you can study their reproduction over at least two generations. Many biologists, for example those who work on preserved specimens, are unable to study the reproduction of an organism, so they use the second group of criteria.

✓*Quick check 3 and 4*

Extinction is a fact of life on Earth. It is estimated that life has existed on Earth for 4000 million years. Over that time there have been five mass extinctions, and it is believed that we are now entering another one caused by a dominant life form – ourselves. This makes surveys of global biodiversity of great urgency. At one level this is an inventory – a list of all the species on Earth. No-one knows for sure how many species there are. Often experts disagree about whether similar organisms belong to the same species or to different species. Some experts lump similar organisms together ('lumpers'); others split them into different species ('splitters').

The Catalogue of Life is managed by Species 2000 and the Integrated Taxonomic Information System (ITIS), and is planned to become a comprehensive catalogue of all known species of organisms on Earth by the year 2011. The list for 2007 contains 1 008 965 species. This is thought to be just more than half of the world's known species. Estimates of global **diversity** are constantly being revised upwards as scientists look more carefully at species and discover new ones. Tropical forests and the deep ocean are two areas that constantly provide examples of new species.

It is important that we identify, name and document what exists on Earth before it disappears. Knowing and understanding the species that exist, we can perhaps do something to help conserve them and prevent their extinction. But just listing species is only one aspect of biodiversity – it does not take into account the number of individuals of each species, the genetic variation within each species, or the range of habitats and ecosystems that they inhabit.

✓Quick check 5 and 6

QUICK CHECK QUESTIONS

1 Define the term biodiversity. Explain how biodiversity is assessed.
2 Define the term *habitat*, and give three examples.
3 Define the term *species*, and explain why it is not always possible to apply this definition to some species.
4 Explain why a biologist would need to study two generations to determine whether a recently discovered organism is a new biospecies.
5 Explain why it is important to conserve the biodiversity of planet Earth.
6 Explain why a list of species is not a complete record of biodiversity.

Sampling

Key words

- sampling
- abundance
- distribution
- transect
- quadrat
- Simpson's diversity index

When measuring the biodiversity of a habitat, you need to find out:

- what species are present
- the **abundance** of each species
- the **distribution** of each species across the area.

The first thing to do is compile a species list. You should search hard for different species and use an identification key to name what you find. Some keys are dichotomous keys, although there are other designs, such as picture keys.

Plants are fairly easy to find, but not always so easy to identify. Animals are less easy to find and it may be necessary to set traps to catch them, especially those that are nocturnal. Pitfall traps catch nocturnal ground beetles, for example. The next step is to measure distribution and abundance. Distribution is where the species are found; abundance is how many of each species are present. As it is impossible to count every organism in a habitat, you have to take representative samples of the habitat and then multiply up to make an estimate of abundance.

Module 3

Quick check 1

Hint

A species list is not a complete measure of the biodiversity in a habitat. Look at the definition of *biodiversity* on page 72.

Techniques for sampling a habitat

Line transects are a good way of showing the distribution of species across a habitat. A tape measure is put across the habitat. All the species touching the line are recorded. The start position of a line transect should be chosen randomly and then the line placed across the habitat. This is a useful technique for showing how the distribution of species changes, for example across a rocky shore or a sand dune. It is not a quantitative method, but gives an immediate impression of the habitat.

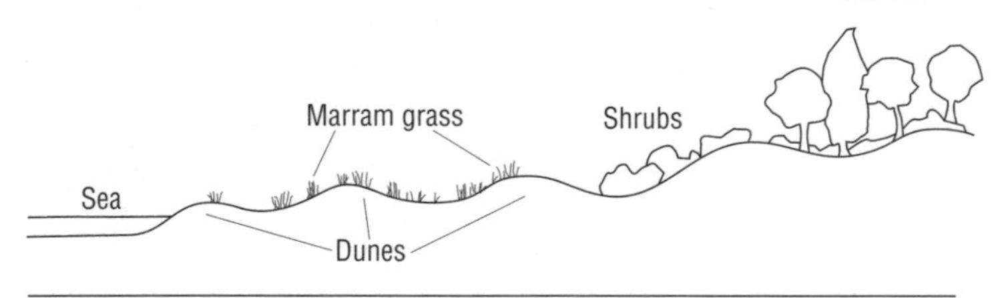

A line transect across a sand dune. This can show the distribution of marram grass – a xerophyte that is adapted to survive in dry conditions

Organisms are distributed unevenly in most environments, so random sampling is used to determine the number and abundance of species present. Random sampling also eliminates any bias on the part of the person taking the samples. The area to be sampled is mapped and given a grid of numbered squares appropriate to the sampling technique. The squares to be sampled are determined by the use of random number tables or random numbers generated by a calculator or a computer.

Quick check 2

Quadrats are square metal or plastic frames used for taking quantitative results. They come in different sizes, and some are subdivided into smaller squares. A suitably sized quadrat is chosen for the habitat being investigated, and placed on the ground at the coordinates generated by using random numbers. The abundance of each species may be determined in one of several ways.

- **Percentage cover** = the proportion of the quadrat's area occupied by the species.
- **Population density** = the number of individuals per quadrat.
- **Species frequency** = the proportion of quadrats with the species present.
- An estimate of abundance using a subjective scale, such as **ACFOR** (abundant, common, frequent, occasional and rare).

Hint

In a survey of a lawn, 15 out of 20 quadrats contained dandelions. The species frequency is 0.75.

Species richness is the number of species present in the study area. The larger the number of different species counted, the greater the species richness. This may be misleading as there may be one dominant species (with many individuals) and some species with one or two individuals.

Species evenness takes into account the number of individuals of each species present in an area.

With an increase in species richness and species evenness, there is an increase in biodiversity. Simpson's diversity index is a measure of biodiversity that takes into account these two aspects of diversity.

Using a quadrat to estimate percentage cover. Each small square represents 1% of the total area. Where a plant occupies part of the square, you have to estimate the cover.

Simpson's diversity index

A group of students used nets to take samples of the animals in a stream. They counted all the individuals of each species that they found and collected the results in a table.

✓Quick check 3 and 4

Species	Number of individuals (n)	Number of individuals of each species (n)/total number of individuals (N)	$(n/N)^2$
Freshwater shrimp, *Gammarus pulex*	150	0.289	0.084
Water hog louse, *Asellus aquaticus*	32	0.062	0.004
Mayfly nymph, *Baetis rhodani*	113	0.218	0.047
Wandering snail, *Lymnaea peregra*	2	0.004	0.000016
Caseless caddis fly nymph, *Rhyacophila sp.*	12	0.023	0.000529
Midge larva, *Chironimidae*	210	0.405	0.164
Total	519 (=N)		0.299

The formula for calculating the diversity index is:

$$D = 1 - \Sigma(n/N)^2$$

So here the diversity index is calculated as:

$$D = 1 - 0.299 = 0.7$$

An index near 0 means that there is little diversity. A value near 1 means that there is a high diversity. The index is the probability that any two individuals taken at random from the sample will belong to different species. An index of 0.9 means that the probability is very high; an index of 0.1 means it is low. An ecosystem with a high diversity index is likely to be more stable than one with a low index, so more likely to withstand changes such as pollution than an ecosystem with low diversity.

Hint

The symbol Σ means 'the sum of'.

✓Quick check 5

QUICK CHECK QUESTIONS

1 Why do biologists take samples when studying habitats?
2 Explain why it is important to use random sampling techniques when studying a habitat.
3 Explain how to use a quadrat to find the percentage cover of marram grass on a sand dune.
4 Discuss the advantages and disadvantages of using line transects and quadrats for assessing biodiversity.
5 Ecosystems with high diversity are more likely to withstand changes. Suggest some changes, other than pollution, that may happen to them.

Classification

Key words

- classification
- taxonomy
- taxon
- phylogeny
- binomial system

Classification is the grouping of organisms into categories based on various features. The study of the principles of classification is **taxonomy**. This involves studying the features that are used in classification and how to use them in designing a classification. Each classificatory group (**species**, **genus**, **family**, etc.) is called a **taxon** (plural: taxa). The species is the fundamental group in the hierarchical scheme. Species are classified into groups that show many similarities of features, such as morphology, anatomy, physiology, behaviour and ecology. These groups of species are classified in turn into fewer, larger groups that share fundamental similarities.

The vertebrates are classified in the **phylum** Chordata. The Vertebrata is a sub-phylum of the Chordata.

The table shows eight taxa of the hierarchical classification system.

Taxon	Example	Number of similarities	Size of group	Degree of relatedness
Domain	Eukaryota	Smallest number of similarities	Largest group	Distantly share a common ancestor
Kingdom	Animalia	↑	↑	↑
Phylum	Chordata			
Class	Mammalia			
Order	Carnivora			
Family	Felidae			
Genus	*Panthera*	↓	↓	↓
Species	*tigris*	Largest number of similarities	Smallest group	Share a common ancestor increasingly recently

Taxonomists identify important similarities and differences between species. These result from the way in which organisms have evolved. The history of an organism's **evolution** is its **phylogeny**. We cannot re-run history to find out how different species evolved, but we can use the characteristics we observe and measure to determine how closely related different species are to one another. The figure shows the evolutionary history of vertebrates. Phylogenies also include fossil forms.

✔Quick check 1, 2, 3, 4

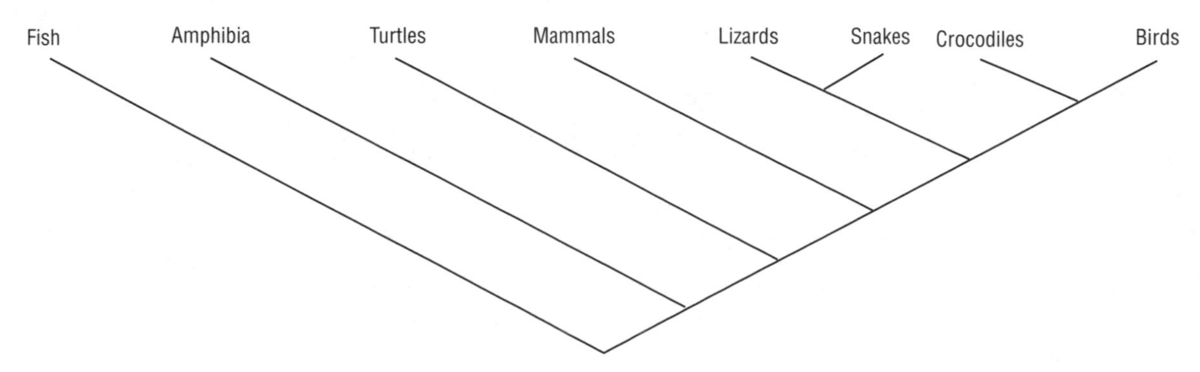

A phylogeny of the main vertebrate groups. This shows that they all come from a common ancestor, and that lizards and snakes are closely related, as are crocodiles and birds.

The binomial system of naming

The table on page 76 shows how to classify the tiger in the hierarchical scheme. Its scientific name is made up of its generic and species names. This is the **binomial system** (binomial=two names): *Panthera* is the generic name (for the genus) and *tigris* is the species name. Both are derived from Greek words. Many new species are discovered and named every year. The names must follow international rules of nomenclature (naming) which include giving them in Latin. The Calayan rail is a bird that was discovered by scientists on Calayan Island in the Philippines in 2004. It was named *Gallirallus calayanensis*, as it is related to other rail species and so was classified in the genus *Gallirallus*. Its species name means 'of Calayan'.

✓ *Quick check 5*

Dichotomous keys

How do you identify an organism? The usual way is to use a **dichotomous key**. This asks a series of questions about the organisms concerned. For each question there are two answers (hence *di*chotomous). When you have answered each question you are directed to another question or to an identification. This is an example of a dichotomous key to some invertebrates that are found in freshwater streams.

Key to some freshwater invertebrates

1	Animal with a shell	Go to 2
	Animal without a shell	Go to 3
2	Shell is a flat coil	*Planorbis corneus*, great ramshorn snail
	Shell is a spiral	*Lymnaea peregra*, wandering snail
3	Body is segmented	Go to 4
	Body is not segmented	*Polycelis felina*, flatworm
4	No legs	Go to 8
	Jointed legs	Go to 5
5	Three pairs of legs	Go to 6
	More than three pairs of legs	Go to 7
6	Body flattened sideways	*Gammarus pulex*, freshwater shrimp
	Body flattened downwards, looks like a woodlouse	*Asellus aquaticus*, water hog louse
7	Two hooks on last segment	*Rhyacophila sp.*, caseless caddis fly nymph
	Three tail appendages	*Baetis rhodani*, mayfly nymph
8	Long breathing tube at end of body	*Eristalis sp.*, rat-tailed maggot
	No long breathing tube	Go to 9
9	Thin body, no suckers	Oligochaete worm
	Suckers at both ends	Leech

Examiner tip

Use a variety of keys for different groups of organisms or for different habitats.

Hint

You can find photographs or drawings of these animals and check that the key works!

QUICK CHECK QUESTIONS

1. Define the terms *classification*, *taxonomy* and *phylogeny*.
2. Explain the link between classification and evolutionary relationships.
3. Explain the term *hierarchical classification*.
4. List the hierarchy of taxa used in biological classification.
5. A new species is discovered and will be given a scientific name. Describe the way in which it will be named.

The five kingdoms

Module 3

Key words

- kingdom
- domain
- homologous
- DNA sequencing

How many kingdoms?

A major problem in taxonomy is how many **kingdoms** (the largest taxon) to use to group living organisms.

In the eighteenth century, the Swedish scientist Carl Linnaeus developed a system of classification in which living things were divided into two kingdoms – the animal kingdom and the plant kingdom. Scientists continued to use this two kingdom system until the second half of the twentieth century. Advances in biology showed that this was not a satisfactory classification. For example, bacteria, algae and fungi are significantly different from green plants, such as mosses, ferns, conifers and flowering plants. Difficulties such as this were solved by having five kingdoms as shown in the table.

Kingdom	Features	Examples
Prokaryotae (Monera)	Cells lack nuclei organised within nuclear envelopes, membrane-bound organelles and microtubules; DNA is circular	Bacteria, including photosynthetic cyanobacteria (blue-greens)
Protoctista	Eukaryotes; one-celled or assemblages of similar cells; some are autotrophic, some heterotrophs	*Stentor*, *Amoeba*, algae (including seaweeds), slime moulds and all eukaryotic organisms that are not fungi, plants or animals
Fungi	Eukaryotes with cell walls of chitin rather than cellulose; they are organised into multinucleate hyphae and are non-photosynthetic (heterotrophs) with absorptive methods of feeding	Mushrooms, moulds such as *Penicillium* and yeasts
Plantae	Eukaryotic organisms that are multicellular photosynthetic autotrophs and have cellulose cell walls; plants are non-motile	Liverworts, mosses, ferns, conifers and flowering plants
Animalia	Eukaryotic organisms that are multicellular and non-photosynthetic (heterotrophs) with nervous coordination	Jellyfish, coral, worms, insects, spiders, fish, amphibians, reptiles, birds, mammals

Quick check 1

Note that the viruses are not fitted into this five-kingdom classification. They do not have a cellular structure and do not carry out any life processes on their own. They are classified into the group Akaryotae, and further subdivisions are based on the structure of their genetic material – DNA or RNA.

Hint

The terms *autotroph* and *heterotroph* are explained on page 62.

Methods of classification

Biologists in the nineteenth and early twentieth centuries constructed classification systems using the features that they could see with the naked eye and with the light microscopes of the time. Classifications were based on observable features including:

- external appearance (morphology)
- internal structure (anatomy)
- development (embryology).

Taxonomists searched for features that were **homologous**. These are features that are thought to have an evolutionary origin in the same ancestral structure. An example is the pentadactyl limbs of the different tetrapod (four-limbed) vertebrates: amphibians, reptiles, birds and mammals.

Quick check 2

Scientific developments have revolutionised biology over the past 60 years. Now many other features are used by taxonomists, who use databases to store, search and organise huge amounts of information.

Protein primary structure

Sequencing the **primary structure** of proteins shows that closely related organisms have sequences that are nearly identical. The sequences of unrelated organisms have greater differences, although there may be similarities – which is not surprising if the protein has the same function in these organisms. The table opposite shows the number of amino acids that differ from those in the β-polypeptide of human haemoglobin, which is made of 146 amino acids, as are most of the others in the table. The last three organisms have molecules that perform the same function but do not have α- and β-polypeptides.

Soya beans have root nodules full of nitrogen-fixing bacteria. Leghaemoglobin combines with oxygen to restrict the oxygen available in the root nodules. This allows the bacteria to fix nitrogen successfully. Although the primary sequence is very different from the human β-polypeptide, the shape of the molecule is very similar – not surprising, as they both contain haem and have the same function.

Species	Number of amino acids that differ from human β-polypeptide
Human	0
Gorilla	1
Rhesus monkey	8
Dog	15
Horse, cow	25
Chicken	45
Frog	67
Lamprey (a fish-like animal)	125
Sea slug (a mollusc)	127
Soya bean (a legume)	124

Adapted with permission from Kimball's Biology Pages (http://Biology-Pages.info)

Scanning electron microscopy

The scanning electron microscope has allowed scientists to look in great detail at morphology. Surface structures invisible to the naked eye, and often indistinct in a light microscope, can be studied in great detail. For example, rhododendrons have small scales on their leaves. Study of these reveals details that help confirm the classification of these plants. The taxonomy of flies relies on data obtained from scanning electron microscopy of bristles on the fly's body.

Hint

Haemoglobin structure is described on page 42.

DNA sequencing

As classification systems should reflect phylogeny, it seems that the best evidence to use should come from **DNA sequencing**. A comparison of the DNA of different organisms should reveal how closely related they are. Nucleotide sequence data and protein sequences are collated by the National Center for Biotechnology Information in the USA, which makes such information widely available.

Hint

Protein and DNA sequencing provide evidence for evolution.

Three domains

The study of bacteria that live in extreme environments led biologists to realise that these extremophiles share many features with eukaryotes. Study of genes that code for **ribosomal RNA** led to the idea that two groups of bacteria and all eukaryotes had separate origins, and these groups were given the taxonomic status of **domains** to reflect this. Other evidence from flagella structure, membrane structure and details of protein synthesis supported this. The three domains are:

- Archaea
- Bacteria
- Eukaryota.

These are above the level of kingdom, so that each domain is subdivided into kingdoms, phyla, etc. as in the table on page 76.

✓Quick check 3 and 4

QUICK CHECK QUESTIONS

1 Explain why the viruses cannot be fitted into the five-kingdom classification.
2 Define the term *homologous* in the context of taxonomy.
3 Describe the types of evidence that taxonomists use to make their classifications.
4 What is the three-domain classification, and why was it introduced?

Variation and adaptation

Key words

- variation
- discontinuous variation
- continuous variation
- adaptation

✓Quick check 1

Variation

Variation refers to the differences that exist among organisms. Some of this variation we can see; some is revealed only when we study the anatomy, cell structure, physiology, biochemistry and molecular biology of organisms.

Look at domesticated cats and dogs. You can hardly confuse them. There are many different breeds of cats and dogs, all with their breed characteristics. But all dogs belong to the species *Canis familiaris* and all cats to *Felis silvestris*. Cats and dogs are examples of:

- variation *between* species – the differences that are used to assign them to different species
- variation *within* species – e.g. the differences between the different breeds.

✓Quick check 2

In both cases, we can find features to categorise and measure. Differences between species are used in taxonomy to help build up classification systems, as we have seen.

Variation within species

Within a species there is a great deal of variation. Look closely at people around you – there are differences that you can see, but there are also all the biochemical and physiological differences that you cannot see. Some aspects of variation are of the either/or type – this is known as **discontinuous variation**. Examples are the shape of earlobes (attached and unattached), and blood groups. **Continuous variation** is the existence of a range of types between two extremes, human height being a good example.

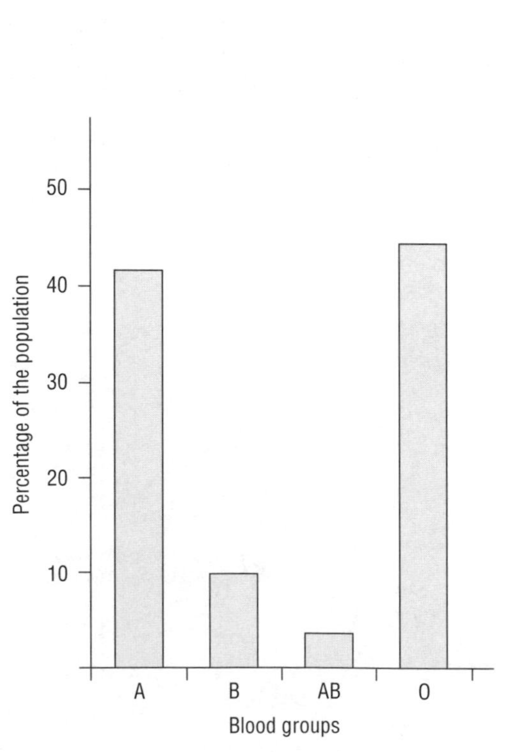

Bar charts, like this one showing the frequencies of the four blood groups in the UK, are used to present data showing discontinuous variation

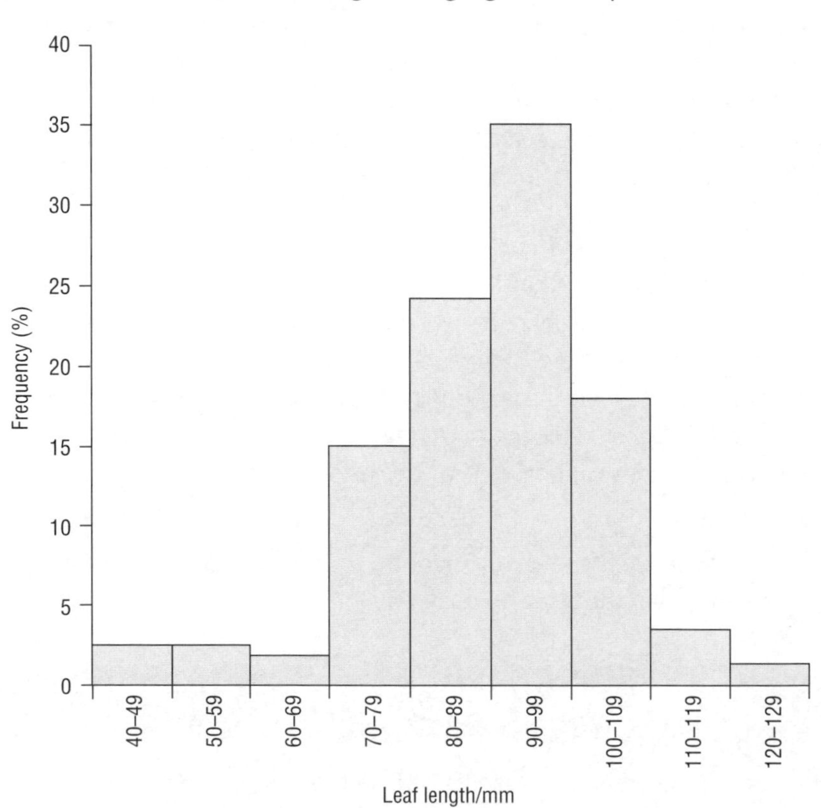

Frequency histograms, like this one showing the lengths of leaves from a bay tree, are used to present data showing continuous variation

Module 3

Discontinuous variation is variation where a feature is one thing or another – there are no intermediates. Some species of flowers may be different colours. Some bacteria of the species *Salmonella typhimurium*, which causes gastroenteritis, are resistant to one of the drugs used to control them, and some are not. A good example in humans is your blood group – A, B, AB and O (see graph opposite). These features are determined genetically and the environment has no (or little) effect.

Genes and the environment interact in controlling features that show continuous variation. This is variation that you can measure. If you collect leaves from a tree and measure their lengths, you will find a range from longest to shortest, and can calculate a mean length. To display this information you would divide up the range into groups and use a tally chart to record the number of leaves in each group. The data are plotted as a histogram.

Hint

See page 62 for examples of the genetic and environmental influences on variation.

✓Quick check 3

Adaptations to the environment

Organisms must fit in to their environments if they are to survive. This 'fitting in' is called **adaptation**. Organisms show different types of adaptations to their environment and way of life: structural, physiological, biochemical and behavioural. These are features that have evolved over time and are continually subjected to **selection pressures**. The table shows three types of adaptations:

- *Rhizobium*, the bacterium that fixes nitrogen and lives in symbiosis with legumes
- Venus flytrap, which grows in bog habitats where there is little available nitrogen as nitrate
- fennec fox, which lives in the Sahara.

Hint

See how many other examples of adaptation you can find within this book.

Adaptation	Microorganism: *Rhizobium*	Plant: Venus flytrap	Animal: fennec fox
Structural	Contains the enzyme nitrogenase, which fixes nitrogen	Some leaves form traps	Large ears
Behavioural	Moves through soil towards legumes (see hint below)	Insects trigger sensitive hairs that stimulate trap to close	Stays in a burrow during the day, emerges at night
Physiological	Exchanges amino acids for carbon and energy sources from the host plant	Leaves secrete extracellular enzymes that digest insects	Ears have large surface area to lose heat, also give animal good direction-finding ability when hunting

Hint

Rhizobium lives inside the root nodules of legumes. See page 79 for an example of an adaptation of soya bean to give *Rhizobium* a suitable environment to fix nitrogen.

✓Quick check 4 and 5

QUICK CHECK QUESTIONS

1. Define the term *variation*.
2. Explain the difference between variation *between* species and variation *within* species.
3. What are the differences between discontinuous and continuous variation?
4. Explain the term *adaptation*.
5. Make a table similar to the one above, and use it to compare the anatomical, behavioural and physiological adaptations of three other species: a plant, an animal and a microorganism.

Hint

Look through the rest of this Revision Guide to find your three species when answering quick check question 5.

Natural selection

Key words

- evolution
- natural selection
- speciation

Evolution is the gradual development of organisms over time. Charles Darwin proposed **natural selection** as a mechanism to explain how evolution has occurred. The idea that organisms had evolved was not new, but Darwin was one of the first to suggest how this could have occurred.

Evolution by natural selection

Darwin made four observations:

- organisms tend to produce a large number of offspring
- populations of organisms remain fairly stable over time
- variation exists among offspring, but
- offspring tend to appear similar to their parents, so inherit their features from them.

The consequences of these four observations may be summarised as follows.

Overproduction

All organisms produce far more offspring than are needed to maintain a constant population; for example, a pair of elephants could produce 19 million descendants after 700 years. Huge increases in populations do not usually occur because population numbers are limited by various factors.

Population levels

Animals compete for food, water, territory and breeding sites. **Herbivores** graze on plants and prevent them growing too much; predators check any increase in populations of herbivores. Diseases kill plants and animals. Most microorganisms starve to death, as they rarely find themselves in places where food is available, but when they do they reproduce at rapid rates.

Competition for rare resources means that there is a struggle for existence between members of the same species. They all have the same features, and obtain those resources in the same way. Many plants and animals die young because of starvation, predation or disease. Many never have the chance to reproduce and pass on their alleles.

The physical environment kills many individuals, for example during storms, floods and droughts.

Variation

There is often much variation within species that reproduce sexually. Genetic information is 'shuffled' every time meiosis occurs during gamete formation. **Mutations** give rise to new forms of DNA. Individuals inherit different alleles from their parents. In the struggle for existence, it is the organisms that are the most successful at gaining resources that survive and reproduce. These find sufficient food, escape predation, survive disease, defend a territory and find a mate. These successful organisms are better adapted to their environment than others, which may die before reproducing. Successful plants are those with adaptations that give them the competitive edge in their struggle to absorb sufficient light, minerals and water and defend themselves against grazers, pests and disease. Factors that influence selection are the availability of food and water, predation, disease, competition, and many physical factors, such as temperature.

Inherited features

Organisms that are better adapted to the environment reproduce and leave offspring, so that their alleles are passed on and become more common. These alleles are the ones that give organisms the ability to compete successfully and survive. Less successful alleles, or harmful ones, become rare.

Hint

A **gene** is a length of DNA that codes for one or more polypeptides; genes exist in different forms called **alleles**. Two individuals of the same species have the same genes, but are likely to have different alleles for many of those genes.

✓*Quick check 1*

Hint

You will learn more about mutation at A2.

Populations of the same species may become isolated by geography (mountains, seas, etc.). Selective pressures are likely to be different, so populations will develop with different adaptations. **Speciation** occurs when the two populations are no longer able to breed together to produce fertile offspring.

Hint

Avoid using the term 'survival of the fittest' in an explanation of natural selection. It does not explain anything.

Selection in action

The first antibiotic, penicillin, became widely available during the Second World War in the 1940s. By using antibiotics, we have changed the environment for bacteria such as *Salmonella* and *Mycobacterium*. Any individual bacterium that has a mutation allowing it to overcome the effect of an antibiotic has a selective advantage. It will survive, while those without this feature will be killed or their growth will be inhibited by the antibiotic. Over time, the number of resistant types of bacteria has increased so that antibiotics introduced years ago are now ineffective.

✓Quick check 2 and 3

Hint

Penicillin is an enzyme inhibitor – see page 57.

The insecticide DDT was first widely used during the Second World War to control mosquitoes and other pests. Since then, many chemicals have been developed to control insect vectors such as *Anopheles* mosquitoes, and insect pests of crops and livestock. Pesticides are a selective agent. Individuals with an allele that confers resistance to pesticides are better adapted, and thus will survive and breed.

The consequences of resistance of microorganisms to antibiotics, and of insect pests to insecticides, mean that the pharmaceutical and agrichemical industries are continually searching for new products. This is sometimes compared with being on a treadmill or in an arms race. No sooner has a new product been introduced than resistance appears, necessitating the search for more new products.

Hint

Anopheles is the vector of malaria – see page 64.

Evidence for evolution

Fossils are the remains of organisms that are preserved in sedimentary rocks. Fossils provide evidence for evolution as they show:

- a progression over millions of years from simple bacteria to complex animals and plants
- many structural similarities with living organisms
- stages in the evolution of present-day organisms (e.g. the horse, elephant and human).

Molecular evidence

Many biological molecules are the same in all organisms – examples are DNA, RNA, ATP, proteins, phospholipids, polysaccharides and coenzymes, such as NAD. This argues for a common ancestry for all life on Earth. Although there are many differences between proteins, such as haemoglobin, from different organisms, there are enough similarities to show the relationships between different organisms, as we have seen on page 79.

Hint

Other evidence is protein structure and DNA sequencing – see page 79.

The primary structure of proteins is determined by the sequences of bases in DNA. Gene sequences of closely related species are found to be very similar. Similarities and differences can be used to group species and the extent of the differences gives an idea of when speciation occurred.

QUICK CHECK QUESTIONS

1. State Darwin's four observations and outline how he made arguments leading to the theory of natural selection.
2. Explain how human activities have influenced selection in bacteria and insect pests.
3. Suggest how a population of bacteria became resistant to several antibiotics.

Conservation: maintaining biodiversity

Key words

- conservation
- extinction
- keystone species
- genome

We live in a time of great biodiversity (perhaps the largest ever): tropical forests and coral reefs are two of the most species-rich areas on Earth. We also live in a time when our activities are causing severe problems for many ecosystems and driving many species to **extinction**. **Conservation** is the protection of ecosystems, habitats and species by taking action to halt destruction and extinction. Conservation often involves managing areas of land and taking steps to encourage species in their habitats; in extreme cases it involves removing animals to captivity or growing plants in cultivation.

Reasons for conservation

Economic

Natural ecosystems provide us with many valuable goods and services. The goods are obvious – products such as fish and timber that we harvest from the wild. The services are less obvious, but think of human wastes that are broken down, water that is filtered and purified, soil formation and climate regulation. Researchers estimate that US$250 billion a year is lost because of habitat destruction.

Quick check 1

Ecological

Keystone species play significant roles in the regulation of ecosystems. Forests of kelp (a large seaweed) in the Pacific off the coast of North America began to disappear when sea otters were hunted by humans. Sea otters eat sea urchins. Without predation by otters, the population of sea urchins increased and overgrazed the kelp. Preventing the hunting of sea otters restored the balance. In this ecosystem the sea otters were the keystone species.

Quick check 2

Ethical

Many species have become extinct as a result of human action, for example the great auk, passenger pigeon, dodo and thylacine (Tasmanian tiger – a marsupial). It is estimated that over 20 000 species of animals and plants face extinction in the near future. Humans have a responsibility to maintain species, ecosystems and habitats for future generations.

Aesthetic

Wild places and man-made ecosystems, such as chalk downland and heathland in the UK, are beautiful places that people enjoy. These places should be conserved for future generations. People enjoy observing wildlife, be it big game in East Africa or garden birds in the UK. The large animals and plants that we enjoy seeing in their natural habitats are sustained by a web of interactions that involve huge numbers of species – many of them microorganisms that few would consider to be attractive to look at, even if we could see them!

Biodiversity for agriculture

By maintaining biodiversity, we are retaining a huge reserve of genetic information that we may wish to call on in the future. The human diet is very limited compared with that of our ancestors. Since agriculture developed in the Neolithic Age (10 000 years ago), our diet has become restricted to very few species. (Make a list of the different species of plants, animals and microorganisms that feature in your diet, and you will see what we mean.) If diseases destroyed the three great staple foods – wheat, maize and rice – there would be huge starvation and loss of life. As stocks of familiar fish, such as cod, have decreased, fishermen have hunted fish that previously were considered unpalatable.

There may be plants and animals that we could farm that would reduce our reliance on a small number of species. Wild grasses of Ethiopia and the Middle East are related to wheat and barley, and hold alleles of genes that could be used to vary the **genomes** of our crops that are genetically uniform and susceptible to disease. Rare breeds of animals and landraces of crop plants are examples of the variation that existed in the past; these have largely been replaced by modern varieties produced by selective breeding in the twentieth century (see page 62). It is estimated that locally adapted breeds of animal are lost worldwide at the rate of one a week. These need to be conserved to retain the genetic potential they may have in the future.

Hint

You will learn more about genomes at A2.

Hint

A landrace of a crop plant is a variety that has been selected by farmers for generations and is adapted to local conditions.

Maintaining biodiversity will be important as the climate changes. Some areas may become too hot and dry to support agriculture; other areas, where crops do not grow now, may become available as the climate becomes warmer – but the local conditions may not suit modern varieties of crops. To breed new varieties that can survive in those conditions, scientists will look at the variation that already exists within crops and their wild relatives.

As the Earth heats up, insect vectors of tropical diseases are likely to spread; examples are *Anopheles* mosquitoes, and vectors of sleeping sickness, dengue fever and yellow fever. Diseases of animals and plants may also spread more easily. Climate change is thought to be the underlying cause of an epidemic of bird malaria in Hawaii that killed thousands of birds, and the spread of an insect-borne viral pathogen that caused distemper among lions in the Serengeti National Park in Tanzania. Crop plants are also at risk. A new strain of a fungal pathogen of wheat, black stem rust, first noticed in 1999, has spread throughout East Africa and threatens wheat-growing areas in India, the USA and Australia. Wheat plants have no resistance to this strain, and the effect could be loss in yield. The disease could spread even further with global warming.

Biodiversity for medicine

Plants have complex metabolisms that produce a very wide range of chemicals, some of which may have potential as medicines. We have seen examples of this: Taxol™ and vinblastine were originally extracted from plants (see page 72). Scientists screen compounds from plants – especially those compounds used in traditional medicines. Currently much interest is being shown in plants used in traditional Chinese medicine. There may be many potential drugs in plants that have yet to be screened. This is why it is important to maintain plant diversity – as well as maintain the diversity of animals and microorganisms

✓Quick check 3 and 4

QUICK CHECK QUESTIONS

1 Discuss the economic reasons for conservation.
2 Explain the term *keystone species* and outline the ecological reasons for conservation.
3 Why is it important to conserve biodiversity?
4 Outline the threats to agriculture and human health posed by climate change.

Conserving endangered species

Key words

- endangered species
- conservation *ex situ*
- conservation *in situ*

✓Quick check 1

Endangered species are those that have such small numbers that they are at risk of extinction. There is little genetic variability within such species; there is a high probability that they will die out as they are susceptible to genetic and infectious diseases.

Zoos and botanic gardens

Conservation *ex situ* happens in places removed from the natural habitat of the species concerned. Zoos cooperate in the captive breeding of endangered species, and specialise in particular species. These schemes are often criticised for:

- removing animals from their environment
- high death rates that occur in captivity
- failure of captive animals to breed
- difficulties in reintroducing captive-bred animals to the wild because they have lost their fear of humans and their immunity to parasites and pathogens, and are no longer adapted to wild conditions.

Examiner tip

Techniques such as artificial insemination, *in vitro* fertilisation, frozen embryos and cloning are all used to help endangered species.

Zoos counter these criticisms by saying that some species would become extinct if left in the wild, as it is too difficult to protect their environment. Some captive breeding programmes have been so successful that there is now a shortage of habitat to receive the animals as is the case with lemurs that originate from Madagascar.

Hint

Not all captive breeding projects involve large mammals and birds.

A new initiative is to collect genetic material (e.g. embryos) from endangered species and store it in a frozen state. When or if animals become extinct there is the possibility of using this material for reintroduction assuming suitable habitats are available.

Captive breeding programmes are linked to conservation projects to create sufficient habitat for reintroductions. The British field cricket, *Gryllus campestris*, nearly became extinct in the 1990s. Captive breeding and release of over 14 000 individuals to selected sites on the South Downs has saved this insect.

The Millennium Seed Bank at Wakehurst Place in West Sussex is the largest *ex situ* conservation project of its kind. It intends to collect and store seed from 10% of the world's plants by 2010. Seeds are kept in cold store and checked periodically to make sure they are viable. These can be used as a genetic resource for future scientists looking for useful genes. It is a botanist's Noah's Ark of species that could become extinct with climate change and habitat destruction. Plants may have a variety of uses in the future, for example in land reclamation following habitat degradation. Some may provide new medicines (see page 85).

Nature reserves

Conservation *in situ* occurs in natural environments and is considered to be far more effective, and cheaper, than *ex situ* conservation.

- National Parks in South Africa and East Africa protect the world's largest land animals.
- National Nature Reserves in the UK protect species such as the snake's head fritillary, *Fritillaria meleagris*.
- Marine reserves, such as the Skomer Marine Nature Reserve off the coast of Pembrokeshire, protect vulnerable marine habitats.

✓Quick check 2, 3, 4

Some threats to wildlife are so extensive that they require international cooperation.

The Convention on International Trade in Endangered Species of Wild Fauna and Flora (**CITES**) is an international treaty that restricts trade in endangered species and their products. Over 30 000 species are protected by being placed on one of three Appendices. Appendix 1 includes species that are threatened with extinction, such as gorillas, tigers, leopards and the Asiatic lion. Plants include the monkey puzzle tree, *Araucaria araucana*; the cycad, *Cycas beddomei*; and the pitcher plant, *Nepenthes rajah*. There is a huge trade in animals, plants and their products, and much smuggling of organisms covered by CITES. Law enforcement in this area is difficult and poaching and smuggling remain severe threats to wildlife worldwide.

The Convention on Biological Diversity (**CBD**) was signed in 1992 in Rio de Janeiro at the so-called Earth Summit. It covers the use and conservation of biodiversity, requiring countries to develop and implement strategies for sustainable use and protection of biodiversity. In response, the UK government launched its Biodiversity Action Plan in 1994; as of 2007 this has action plans for:

- 391 species, including *Gryllus campestris*
- 45 habitats, including chalk downland.

There are also 162 Local Biodiversity Action Plans, including one for Sussex, which includes habitats of *G. campestris*.

Agenda 21 (part of the CBD) promotes the use of **environmental impact assessments**. Ecologists sample an area subject to development (housing, roads, airports, dams) and report on the likely impact on species and their habitats. Developers and planners must take into account the effects of new developments on the environment, and seek to minimise these effects.

✓ *Quick check 5 and 6*

QUICK CHECK QUESTIONS

1. What makes a species endangered?
2. Explain why *in situ* conservation is preferable to *ex situ* conservation.
3. Suggest some circumstances that make *ex situ* conservation necessary.
4. Suggest the disadvantages of maintaining very small populations of endangered species in wildlife reserves.
5. Explain why the UK government set up its Biodiversity Action Plan.
6. What are environmental impact assessments and how are they used?

End-of-unit questions

1 The quantities of the nitrogenous bases adenine, guanine, cytosine, thymine and uracil were measured in samples of different polynucleotides. The margin of error for the measurements was estimated as ±1%. The data are shown in the table.

Source and type of polynucleotide	Adenine	Guanine	Cytosine	Thymine	Uracil
Pig liver DNA	29.4	20.5	20.5	29.7	0.0
Pig spleen DNA		20.4	20.8	29.2	0.0
Pig liver RNA	24.5	29.5	40.8	0.0	15.2
Wheat DNA	27.3	22.8	22.7	27.1	0.0
Wheat RNA	23.6	33.3	27.8	0.0	15.3

(a) Name *two* other components of DNA structure, apart from nitrogenous bases. (2)

(b) The adenine measurement for pig spleen DNA has been omitted. What would you expect this measurement to be? Explain your answer. (2)

(c) Describe the pattern of the data for thymine and uracil in these polynucleotides. (1)

(d) The percentages of guanine and cytosine for both pig and wheat DNA are very different from the percentage in RNA from the same tissues. Suggest why. (3)

(e) Wheat root-tip cells have twice as much DNA as is found in wheat pollen. Explain why. (2)

2 Beta-galactosidase is an enzyme. It acts to hydrolyse a substrate called ONPG, producing a monosaccharide sugar called galactose and nitrophenol, which is a yellow substance.

ONPG (colourless) $\xrightarrow{\beta\text{-galactosidase}}$ **nitrophenol (yellow) + galactose**

substrate **products**

The activity of this enzyme is assessed by the rate at which the yellow colour appears. Two different batches of β-galactosidase solution were obtained: one extracted from yeast and the other from a fungus called *Aspergillus*. A student decided to compare the activity of these two batches of enzyme at different pH.

(a) Outline a plan for carrying out this investigation, stating which variables would need to be controlled to make the procedure valid. (8)

(b) Explain how changes in pH affect the action of enzymes. (2)

(c) When there is a high concentration of ONPG, and the pH and temperature are both optimum, galactose inhibits β-galactosidase. Suggest how you think this happens. (2)

3 The incidence of a disease is the number of new cases that are reported over a certain period of time, such as a week, month or year. Diet is linked to differing incidences of coronary heart disease (CHD).

(a) (i) Name *two* components of diet *other than lipids* which are *positively* correlated with CHD. (2)

(ii) Name *two* components of diets that are *negatively* correlated with CHD. (2)

Examiner tip

Note what you are told about the error margin in the introduction to the question.

Examiner tip

Question 2a asks for an *outline* so you don't need as much practical detail as for a full investigation plan. There are eight marks, so you must make at least eight different points in your answer.

Examiner tip

Anticipate questions on practical biology in your examination papers. Make sure you know the important practical details to use in your answers.

Examiner tip

In answering question 2c, read over the section about enzyme inhibitors. You are being asked to suggest a way in which this **inhibition** occurs.

Examiner tip

Remember that correlation is one of the mathematical ideas which the specification requires you to understand.

Environmental factors, including diet, change the concentration of high-density lipoproteins (HDLs) and low-density lipoproteins (LDLs) in blood plasma.

(b) Explain the association between LDLs and CHD. (4)

People in France tend to eat lots of fatty meat and cheeses with high fat content. The incidence of CHD in France is relatively low.

(c) Suggest why this does *not* disprove the hypothesis that a diet rich in fat increases the risk of CHD. (3)

4 New motorways are being constructed in Poland. It is important to plan the route taken by a new motorway so that the resulting environmental damage is minimised. Poland has large areas of ancient forest. Environmental consultants are employed to determine the likely impact of choosing different routes through these forested areas.

(a) Explain the terms *habitat biodiversity* and *species biodiversity* as they would be used when planning the route of a new motorway. (4)

Most European bison alive today are descended from small populations that survived in Polish forests. Bison have a low genetic diversity when compared with domestic cattle.

(b) Explain the term *genetic diversity*. (2)

Sampling techniques are used to assess biodiversity. One survey used 20 quadrats, selected using a map (divided into a grid) of the proposed motorway route and a table of random numbers. Another survey of the same area selected the 20 quadrats by allowing an experienced biologist to fly across the area in a helicopter.

(c) Apart from cost, discuss the advantages and disadvantages of these alternative sampling strategies. (3)

The birds seen in a quadrat measuring 50 × 50 m were counted each day, between 9.00 and 9.30 am, for 1 week. The total numbers of birds of each species are recorded in the table.

Bird species	Total number of each species seen	Species total/ grand total (n/N)	$(n/N)^2$
Black grouse	2	0.035	0.012
Capercaillie	2	0.035	0.012
White-backed woodpecker	3	0.053	0.003
Three-toed woodpecker	5	0.088	0.008
Red-breasted flycatcher	12	?	0.044
Blackbird	15	0.263	0.069
Robin	6	0.105	0.011
Coal tit	5	0.088	0.008
Crossbill	7	0.123	0.015
Grand total of all birds seen	57	Simpson's diversity index =	

(d) Use the table to:

(i) calculate the missing item in the third column and the Simpson's diversity index. (2)

(ii) explain the term *species richness* as it would be used in this survey. (1)

(e) Suggest why the species totals shown in the table may be misleading. (4)

In Portugal, much ancient forest dominated by native cork oak trees has been replaced by plantations of Australian *Eucalyptus* trees.

(f) Suggest why a *Eucalyptus* plantation would have a lower Simpson's diversity index for birds than would a cork oak forest in the same part of Portugal. (3)

Examiner tip

Questions 4a and 4b give you contexts, which may help you remember these definitions.

Examiner tip

In question 4e you need to use the information about how the data were collected as well as the table.

Examiner tip

Suggest questions usually have several possible answers. You are not expected to think of all of them.

Answers to quick check questions

Many of the quick check questions are designed to test your understanding and recall of material in the spreads. You will find many of the answers quite easily. However, some are more difficult and require you to put several ideas together, and maybe to look at more than one spread when asked to make comparisons. Answers to quick check questions are given here.

Unit 1 – Cells, exchange and transport

Module 1 – Cells

Studying and measuring cells (page 2)

1 Enlarging the photograph does not increase the detail that can be seen – this is dependent on the resolution, which is limited by the wavelength of light used in the microscope.
2 120 mm.
3 15 µm.
4 Most biological material is transparent; gives contrast to material.
5 High resolution; so that details of cell structure can be seen, e.g. membranes, organelles such as ribosomes, mitochondria and endoplasmic reticulum.

Cells and organelles (page 4)

1 Organelles carry out different functions within the cell. There are many examples in the spread.
2 Both cells: nucleus, nucleolus, Golgi apparatus, mitochondria, RER, SER, ribosomes; (i) only in palisade cell: chloroplasts; (ii) only in pancreatic cell: lysosomes (vacuole in plant cells fulfils same function as lysosomes in animal cells).
3 E.g. eukaryotic cell has: nucleus; mitochondria; Golgi apparatus; endoplasmic reticulum. Eukaryote cell does **not** have plasmids, pili, flagellum or mesosome.

Cell membranes (page 6)

1 Phospholipid bilayer is liquid; proteins are dispersed within the bilayer like moveable mosaic tiles.
2 Phospholipids allow small molecules (water, carbon dioxide, oxygen) and fat-soluble molecules to pass through, but not ions or larger water-soluble molecules. Proteins form channels and carriers to allow movement of ions and polar molecules that cannot pass through the phospholipid bilayer.
3 Cell surface membrane, membranes of nuclear envelope, endoplasmic reticulum, Golgi apparatus, lysosomes, mitochondria and (in plants) chloroplasts and the membrane around the large vacuole.

Exchanges across membranes (page 8)

1 Both molecules are polar, but carbon dioxide is much smaller and can pass between phospholipids, glucose is too large to be able to do this.
2 Active transport uses carrier proteins (not channel proteins); uses energy in the form of ATP; moves substances against their concentration gradient.
3 Mitochondria provide energy for active transport of ions.
4 Diffusion is the net movement of molecules or ions from a region of high concentration to a region of lower concentration. It is an example of passive transport. Osmosis: diffusion of water down a water potential gradient through a partially permeable membrane. Facilitated diffusion: diffusion across a membrane through a protein channel or carrier. Active transport: movement of molecules across a membrane against a concentration gradient using ATP from respiration to provide energy. Endocytosis: enclosing of substances in a vacuole or vesicle made by cell surface membrane to bring substances into a cell. Exocytosis: substances within a vesicle or vacuole are removed from the cell when the vesicle membrane fuses with the cell surface membrane. Energy is used moving vesicles to and from the cell surface membrane.

Cell signalling and investigating cell membranes (page 10)

1 These are water-soluble molecules that cannot cross the cell surface membrane.
2 The drugs have shapes complementary to the shapes of the receptors.
3 (i) Red blood cells do not have a cell wall that can withstand the increased pressure inside the cell. (ii) Water diffuses out of the cells by osmosis. The water moves *down* a water potential gradient.
4 Cell surface and vacuole membranes are disrupted by high temperatures. In hot water the phospholipid bilayer becomes more fluid and proteins denature.

Cell division and mitosis (page 12)

1 Chromatin condenses into chromosomes during prophase; becomes visible as a double structure – with two chromatids joined at centromere; attached to the spindle during metaphase. Chromatids pulled apart in anaphase. Each chromatid (now a chromosome) uncoils in telophase. When mitosis is completed, replication occurs during interphase of the next cell cycle to form two molecules of DNA, which condense in the next prophase to form two chromatids.
2 So each chromosome has two identical chromatids to transfer to the two daughter cells.
3 The quantity of DNA doubles in (i) and halves in (ii).
4 The number stays the same.

Chromosomes and meiosis (page 14)

1 Chromosomes that are the same size and shape, have their centromeres in the same place and have the same genes (although they may have different alleles of those genes).
2

Feature	Mitosis	Meiosis
Number of divisions of the nucleus	1	2
Chromosome number in daughter nuclei	same as the parent nucleus	half the number in the parent nucleus
Homologous chromosomes pair up	no	yes
Role in life cycle	growth and asexual reproduction	production of gametes in sexual reproduction

3 The chromosome number halves – diploid to haploid.
4 Ensures that chromosome number remains the same from generation to generation; gives rise to variation.

Tissues, organs and organ systems (page 16)

1 E.g. ciliated epithelium: moves mucus in trachea; blood: transport of oxygen; cartilage: support; muscle: movement.
2 Tissue: a group of similar cells carrying out the same function(s); organ: a group of tissues that form a distinct structure and carry out one major function or functions.
3 Xylem: transports water; phloem: transports sugars.
4 Cells, tissues, organs and organ systems carry out different functions within the organism.
5 Some examples: palisade cells rely on root hair cells to absorb water and ions, and xylem vessels to transport water and ions from roots to leaves; they also rely on guard cells to open stomata to allow carbon dioxide to enter so they can photosynthesise; muscle cells rely on red blood cells to deliver oxygen and remove carbon dioxide.

Module 2 – Exchange and transport

Exchange surfaces and the lungs (page 18)

1 Humans have a very small surface-area-to-volume ratio. The surface area is not large enough to permit enough oxygen to enter by diffusion. Protoctists have a very large surface-area-to-volume ratio and there is sufficient surface for the oxygen they need.
2 Five membranes: two for the squamous epithelial cell; two for the endothelial cell of the capillary; one for the red blood cell.
3 Nose/mouth, trachea, bronchus, bronchioles.

Measuring lung activity (page 20)

1 Tidal volume is the volume of air breathed into the lungs in one breath however deep or shallow the breathing. Vital capacity is the maximum volume of air that can be breathed out after taking a deep breath.
2

Feature	Before exercise	After exercise
Mean tidal volume/ dm^3	0.5	2.5
Vital capacity/dm^3	4.15	4.15
Breathing rate/ breaths min^{-1}	12	9
Ventilation rate/ $dm^3\ min^{-1}$	6.0	22.5
Oxygen consumption/ $dm^3\ min^{-1}$	0.2	2.0

Transport in animals (page 22)

1 Distances are too great for substances to move by diffusion, e.g. between body cells and sites of digestion and gas exchange.
2 ×16; ×4375; ×3.5.
3

Feature	Artery	Capillary	Vein
Thickness of wall	Thick	Very thin	Thin
Composition of wall	Smooth muscle, elastic fibres, collagen	Endothelium (one cell thick)	As artery – but much less
Valves	✗	✗	✓
Blood pressure	High	Low	Low
Function	Transport blood from heart; maintain blood pressure	Exchange of substances between blood and tissues	Transport blood to heart; valves prevent backflow

4 E.g. arteries have thick elastic and muscular wall to withstand high blood pressure; capillaries: wall is very thin (one cell thick) so diffusion distance between blood and tissue fluid/cells is short; veins: have semilunar valves to prevent backflow as pressure is low.

The heart and cardiac cycle (page 24)

1 0.8 s; 75 beats per minute.
2 The valve closes at 1 because the pressure of blood in the ventricle is greater than in the atrium; it opens at 4 because the ventricle is empty and the pressure in the atrium is higher.
3 The valve opens at 2 because the pressure of blood in the ventricle is greater than in the aorta; it closes at 3 because the ventricle is empty and the pressure in the aorta is higher than in the ventricle.
4 Information about the pattern of heartbeat and the heart's electrical activity. This may be useful in diagnosing heart conditions.

Blood, tissue fluid and lymph (page 26)

1 ×2429.
2 In this type of question you should refer to both the structural aspect and its function. Full of haemoglobin for transport of oxygen; no organelles, so giving plenty of space for many molecules of haemoglobin; flexible cell surface membrane to allow cell to change shape to move through capillaries; biconcave shape to give large suface-area-to-volume ratio; small so can fit through capillaries.
3

Feature	Red blood cell	Neutrophil	Lymphocyte
Nucleus	✗	✓ (Lobed)	✓
Cytoplasm	Full of haemoglobin	Filled with many lysosomes	Small amount of cytoplasm
Function	Transport oxygen and carbon dioxide	Phagocytosis – ingest and digest foreign material such as bacteria	Secrete antibodies

4 Red blood cells cannot pass through capillary walls; many white cells are made in lymph nodes and enter blood via the lymph.
5 Squamous epithelial cells are thin, so there is a short diffusion pathway for oxygen and carbon dioxide.
6 Note that the question says '... tissue fluid and cells'. This is covered in another spread, but is relevant here.
Oxygen – diffusion (through phospholipid bilayer);

water – osmosis (through phospholipid bilayer and aquaporins); glucose – facilitated diffusion (through channel proteins); ions – some by active transport (through carrier proteins), others by facilitated diffusion (through channel and carrier proteins).

7 The contraction of the left ventricle in the heart.

8 At X the pressure of the blood is high and this causes pressure filtration; at Y the pressure has decreased so water moves into the capillary *down* a water potential gradient – the water potential of the plasma is *lower* than that of tissue fluid because of the higher concentrations of solutes, such as plasma proteins that are too large to leave the capillaries. At X the force exerted by blood pressure is greater than the tendency of water to enter the blood down its water potential gradient.

Haemoglobin and gas transport (page 28)

1 Haemoglobin has four haem groups, each of which can combine with a molecule of oxygen. Carbon dioxide combines with free amino groups (–NH_2) at the ends of the polypeptides in haemoglobin.

2 Temperature; volume of blood; density of red blood cells; pH.

3 Haemoglobin becomes fully loaded with oxygen in the lungs to form oxyhaemoglobin. In tissues, oxyhaemoglobin dissociates to give up oxygen. Quite small changes in partial pressure of oxygen in the tissues lead to a large decrease in oxygen carried by haemoglobin.

4 Use a ruler to answer this question. Put it vertically on the graph at any partial pressure of oxygen in the range 2–6 kPa.
At any partial pressure of oxygen, fetal haemoglobin has a higher percentage saturation with oxygen. This means that oxygen is transferred across the placenta from maternal blood to fetal blood. Fetal haemoglobin combines with oxygen at the same partial pressure that maternal haemoglobin offloads it.

5 Dissociation curve moves to the right with increasing partial pressure of carbon dioxide. Carbon dioxide reacts with water inside red blood cells to form hydrogencarbonate ions and hydrogen ions. Oxyhaemoglobin is sensitive to hydrogen ions and dissociates to release oxygen.

Transport in plants (page 30)

1 By osmosis; down a water potential gradient from the soil water into root hair cells.

2 Into root hair; across root cortex mainly via apoplast pathway along cell walls; via symplast pathway through cells of endodermis; into xylem vessel in root; up to leaves in xylem in root and stem; via apoplast pathway to cell walls of mesophyll cells; evaporates into air space; diffuses through stomata to air outside leaf (see figures on pages 31 and 30).

3 Blocks apoplast pathway through endodermis; ensures water passes *through* the endodermal cells as there is a water potential gradient between the cortex and the tissues inside the endodermis.

Transport in the xylem: transportation (page 32)

1 Hollow and without end or cross-walls to allow uninterrupted flow of water; supported by lignin to prevent inward collapse when rates of transpiration are high and water columns are under tension; waterproofed by lignin to prevent loss of water to surrounding tissues.

2 For example, cut the stem under water to prevent air locks in xylem; leave to adjust to conditions; keep all conditions constant apart from wind speed; take replicates and calculate means.

3 Four points are given on page 33.

Transport in the phloem (page 34)

1

Feature	Xylem vessels	Phloem sieve tubes
Contents	Hollow (no contents)	Some cytoplasm
Cross-walls	None	Sieve plates with sieve pores
Walls	Thickened with cellulose and lignin	Thin, no lignin
Function	Transport water and ions	Transport assimilates (e.g. sucrose and amino acids)
Direction of flow	Upwards	Both directions
Source(s)	Roots	Mature leaves Storage organs (such as roots)
Sink(s)	Stems, leaves, flowers, fruits and seeds	Stems, roots, storage organs, flowers, fruits, seeds, young leaves

Notice that you may use the terms source and sink for xylem as well as phloem.

2 End walls with sieve pores to allow phloem sap to flow freely; no nucleus, little cytoplasm to reduce resistance to flow; cell membrane may prevent loss of sucrose to adjoining cells; sieve plates may prevent sieve tubes from bursting under pressure; plasmodesmata to allow sap to move to and from companion cells.

3 Loading; source is photosynthesising mesophyll cell; sucrose passes to companion cell via plasmodesmata or along cell walls; companion cells pump sucrose into sieve tube element. Sink, e.g. root; companion cells in the root unload sucrose and may convert it into storage substance, e.g. starch.

4 Sucrose loaded at source; lowers water potential; water diffuses into sieve tube; pressure increases, forcing sap out of source; sucrose unloaded at sink; water diffuses out of sieve tubes, lowers pressure. There is a pressure gradient from source to sink.

Unit 2 – Molecules, biodiversity, food and health

Module 1 – Biological molecules

Water and macromolecules (page 38)

1 Water is polar owing to uneven distribution of charge shown on page 38. Hydrogen bonds form between water molecules so they are not free to exist in the gaseous state.

2 As a reactant in hydrolysis reactions; solvent in cytoplasm and body fluids; coolant (e.g. in sweat and breath); provides hydrostatic skeleton in some animals; provides turgidity in plant cells and in the eye.

3 Polysaccharides/carbohydrates – starch; proteins – haemoglobin; lipids – triglycerides; nucleic acids – DNA.

4 C, H, O, N, S, P. (Others mentioned in this book are zinc (Zn) and iron (Fe).)

Proteins (page 40)

1 Draw the diagram shown on page 40 in which a dipeptide is formed from two amino acids; omit the arrow labelled hydrolysis and show a water molecule being removed.
2 Primary – the sequence of amino acids.
Secondary – polypeptide is folded into an α-helix or a β-pleated sheet.
Tertiary – complex folding of the secondary structure to give a specific 3D shape.
3 Bonds between the R groups of amino acids stabilise the tertiary structure; these are hydrogen bonds, disulfide bonds, ionic bonds and hydrophobic interactions. Amino acids that are far apart in the primary sequence are held close together by these bonds in the tertiary structure. This gives rise to specific shapes.
4 Tertiary shapes are specific, so only accept molecules with a complementary shape. Enzymes are specific to their substrates, antibodies to their antigens and receptors to their signalling molecules.
5 Bonds that stabilise tertiary structure are broken; specific tertiary structure is changed (e.g. at active site) so there is a loss of 3D shape.

Globular and fibrous proteins (page 42)

1 1 120 000 000.
2 Haemoglobin is made of more than one polypeptide (it is made of four).
3 Haemoglobin is an intracellular protein. Disulfide bonds do not form in the reducing environment of cytoplasm. Disulfide bonds stabilise extracellular proteins (such as those in the plasma) where there is a strong oxidising environment.
4 Globular protein – complex tertiary/3D structure, usually soluble in water; e.g. haemoglobin. Fibrous proteins have a linear structure and are insoluble in water; e.g. collagen.
5 Collagen molecules are triple helices with many hydrogen bonds between the polypeptides. Strong covalent bonds hold molecules together. Molecules overlap without any lines of weakness.
6

Feature	Haemoglobin	Collagen
Type of protein	Globular	Fibrous
Number of polypeptides	4	3
Shape of molecule	Complex 3D shape	Triple helix
Function	Transports oxygen and carbon dioxide	Strengthens tissues

Carbohydrates (page 44)

1 Monosaccharide: simple sugar that cannot be broken down to form another sugar. Disaccharide: complex sugar formed of two monosaccharides joined by glycosidic bond. Polysaccharide: polymer of many monosaccharide sub-units joined by glycosidic bonds. Glycosidic bond – covalent bond that forms between two sugars by a condensation reaction.
2 During the reaction between –OH groups on carbon atoms 1 and 4, water is formed leaving an oxygen 'bridge' between the two glucose units. A glycosidic bond is broken by the addition of water.
3 These molecules are insoluble, compact and can be broken down into many molecules of glucose.
4 There should be two parts to your diagram. The first part should show at least five glucose monomers at the end of an amylose chain. You should show how water interacts with the glycosidic bond (as in the figure on page 44). Then draw the second part to show three glucose monomers with a free maltose molecule. The enzyme is amylase.
5

Feature	Amylose	Amylopectin	Glycogen	Cellulose
Monomer	α-glucose	α-glucose	α-glucose	β-glucose
Structure	Unbranched, helix	Branched	Very branched	Unbranched, straight chain
Location	Plants	Plants	Animals	Plants
Function	Energy storage	Energy storage	Energy storage	Plant cell walls

6 Cellulose has many projecting –OH groups. These form hydrogen bonds with other molecules of cellulose; these bonds give cellulose great strength to resist the pressure exerted by plant cell vacuoles on cell walls.

Lipids (page 46)

1 Glycerol with three fatty acids attached by ester bonds.
2 Different fatty acids, which can be saturated, monounsaturated or polyunsaturated.
3 Phospholipids have two fatty acids, triglycerides have three; phospholipids have a phosphate group and another hydrophilic group (e.g. choline), triglycerides do not.
4 Hydrophilic 'heads' of phospholipids are attracted to water, but hydrophobic 'tails' are not. Two layers of phospholipids form a bilayer in water with a hydrophobic core from which water is excluded.
5 Component of cell membranes; used to synthesise steroid hormones including vitamin D; liver controls production of cholesterol and packages it into lipoproteins so can be circulated to other tissues and organs; may be deposited in the walls of arteries and lead to atherosclerosis and cardiovascular disease (page 61).

Practical work with biological molecules (page 48)

1 Both are reducing agents / sugars.
2 Make a solution of the plant material. Divide the solution into two samples, **A** and **B**. Boil **A** with Benedict's solution. If positive then confirms that plant material contains reducing sugar. Boil **B** with hydrochloric acid for several minutes; neutralise with an alkali and boil with Benedict's solution. If the plant material contains non-reducing sugars then there will be more precipitate than with **A** as the non-reducing sugars will have been hydrolysed by the acid to give reducing sugars. **B** will then have a higher concentration of reducing sugars than **A**.
3 0.65%.
4 Make a series of concentrations of a protein (10%, 7.5%, 5%, 2.5%, 1%, 0.5%, etc); add the same volume of biuret reagent to each solution and place in colorimeter and take a reading. Draw a graph like that on page 49. Homogenise the seeds by grinding with a pestle and mortar or using a liquidiser; filter; add same volume of biuret reagent to standard volume of solution made from seeds; put in colorimeter; read result from graph.

Nucleic acids – DNA and RNA (page 50)

1 Double helix; two antiparallel polynucleotides; held together by hydrogen bonds between base pairs.
2 Each nucleotide has a base – either purine (A and G) or pyrimidine (T and C). A always pairs with T; C always pairs with G.
3

Feature	DNA	RNA
Type of sugar	Deoxyribose	Ribose
Bases	A, T, C, G.	A, U, C, G.
Size	Very large	Smaller
Number of polynucleotides	2 (double helix)	1 (variety of shapes)

4 DNA is for long-term storage of information. RNA has three different functions to perform during protein synthesis: transferring information from nucleus to the ribosomes; carrying amino acids to ribosomes; being part of the ribosome (the rest is made of protein).

Replication and roles of DNA (page 52)

1 Stages are described on page 52.
2 DNA provides a sequence of bases that code for the primary structure of polypeptides.
3 Nucleus: stores DNA; transcription occurs here. DNA – sequence of bases determines sequence of amino acids in polypeptide; mRNA – copy of gene, travels to ribosome; ribosome – provides sites for assembly of amino acids into polypeptide; tRNA – brings amino acids to ribosome; rough endoplasmic reticulum – site of attachment of ribosomes; polypeptides travel to Golgi apparatus through the space within RER; Golgi apparatus – modifies protein, e.g. by adding sugars to make glycoprotein, packages protein into secretory vesicles.

Enzymes (page 54)

1 Catabolic – molecules are broken down (e.g. by hydrolysis); anabolic – synthesis of larger molecules from smaller ones (e.g. condensation reactions to form polymers such as polypeptides).
2 Lock and key: substrate fits into active site without any change in shape. Induced fit: enzyme changes shape to surround substrate as it enters.
3 Provides an active site where reaction can occur. In the active site, substrate molecules are brought close together so there is more chance of them reacting.
4 Diagram should show two substrate molecules fitting into an active site. The active site could mould around the substrate molecules; the substrate molecules will have shapes complementary to the active site. The substrate molecules should be shown moving into the active site, inside the active site, then leaving as one product molecule.

Factors affecting enzyme activity (page 56)

1 Low temperature, slow rate of reaction as enzymes and substrates have little kinetic energy and do not collide often. Optimum temperature, rate of reaction at highest, greater kinetic energy but enzyme not denatured. High temperature, rate of reaction is nil because enzyme is denatured. Vibrations within molecule break hydrogen and other bonds, tertiary structure destroyed, enzyme is denatured.
2 The pH changes because the hydrogen ion concentration changes. A change in pH from e.g. 7 to 6 is a 10x increase in the concentration of hydrogen ions. The greater concentration of hydrogen ions may cause hydrogen and ionic bonds to break.
3 **(i)** Zinc/copper; **(ii)** haem; **(iii)** NAD.
4 Potassium cyanide inhibits cytochrome oxidase, which is an enzyme of respiration found in mitochondria. As a result it prevents mitochondria producing ATP so there is little energy for cells to carry out activities, such as active transport and protein synthesis.
5 Statins decrease the concentration of cholesterol in the blood. They do this by inhibiting an enzyme in the liver which is involved in making cholesterol.

Experiments with enzymes (page 58)

1 The substrate concentration decreases.
2 At the beginning there is a high concentration of substrate molecules, so that there are many collisions between enzyme molecules and substrate molecules. With time, the concentration of substrate decreases so there are fewer collisions.
3 In both (a) and (b) the reactions would occur faster.

Module 2 – Food and health

Balanced and unbalanced diets (page 60)

1 The effects of eating much less or much more than required; or not eating sufficient nutrients, such as essential amino acids, essential fatty acids, vitamins or minerals.
2 Mass in kg divided by (height in metres)^2.
3 Cholesterol is not water-soluble, so cannot be transported in solution in the plasma.

Food production and preservation (page 62)

1 Breeder chooses feature(s) to improve; selects individuals showing those features; breeds them together; selects from among offspring those showing improvement; repeats procedure for many generations or until tastes change (e.g. fatty meat to lean meat).
2 Yoghurt, cheese, wine/beer/other alcoholic drink, bread, mycoprotein/Quorn™.
3 Food goes 'off': smells bad, looks bad and tastes bad; may be harmful to health. Food may be salted, pickled, frozen, refrigerated, heat-treated or irradiated to preserve it.

Health and disease (page 64)

1 Malaria: *Plasmodium* is transmitted by an insect vector: female *Anopheles* mosquitoes. The mosquito takes a blood meal containing the parasites from an infected person; later the mosquito takes another blood meal from an uninfected person. The infective stage of the parasite is in the mosquito's saliva. TB: *Mycobacterium tuberculosis* is transmitted in droplets breathed or coughed out by an infected person and breathed in by an uninfected person. Transmission in milk and meat from cattle infected with *Mycobacterium bovis*.
2 See the table on page 65.
3 See the table on page 65.

The immune system (page 66)

1 B lymphocytes differentiate into plasma cells and secrete antibodies; T helper lymphocytes release cytokines to stimulate B lymphocytes to divide and differentiate; T killer lymphocytes seek out and kill infected cells. T and B memory cells remain until the next infection of the same strain of the pathogen, when they will respond quickly.
2 The person has memory cells, which will respond very quickly during a subsequent infection of measles. There is only one strain of measles, so the vaccination remains effective throughout life.
3 The graph should be like the figure on page 67, with time in months on the *x*-axis. The primary response should begin just after the first vaccination at 2 months; there should be secondary responses at 3 months and at 4 months, with the concentration going higher each time.

Vaccines and other medicines (page 68)

1 Vaccines contain antigens to stimulate active immunity.
2 Herd immunity by mass vaccination programmes; ring vaccination of people who are at risk from a few infected individuals. These methods eradicated smallpox and will soon eradicate polio.
3 Surveillance – careful monitoring of cases of influenza and the strains of the influenza virus in the population; making vaccines that provide protection against strains predicted to spread each year; vaccinate 'at-risk' groups such as the elderly and young people with conditions such as asthma.
4 Soil microorganisms, especially *Streptomyces*, have provided many antibiotics. There may be species as yet undiscovered that produce chemicals that kill pathogenic organisms or inhibit their growth. Plants may be sources of chemicals that could act against cancers and other diseases.

Smoking and disease (page 70)

1 Gas exchange system: trachea, bronchi and lungs; cardiovascular system: heart and blood vessels.
2 Answer should describe the effects of tar and carcinogens on the trachea, bronchi and lungs.
3 Answer should describe the effects of carbon monoxide and nicotine on the transport of oxygen by haemoglobin, the heart and arteries.
4 These two lines of evidence are summarised on page 71.

Module 3 – Biodiversity and evolution

Biodiversity (page 72)

1 Biodiversity is a measure of the different ecosystems, the number of species, the number of individuals of each species and the genetic variation within each species. It is assessed by surveying an area and making a list of the different species and recording their distribution and abundance within that area. It is also assessed by measuring the genetic diversity within populations.
2 The place where an organism lives. Examples: estuary, pond, coniferous forest, coral reef.
3 A group of organisms that can reproduce to give fertile offspring. The organisms show close similarity in features such as morphological, anatomical and behavioural characteristics.
4 The parent generation will breed together to give the first generation. To check that these are fertile, they will have to be bred together, or with other individuals of the same type, to give the second generation.
5 The future of our species depends upon it! Other reasons are given on page 84.
6 Biodiversity also includes habitats, number of individuals in each species and genetic variation within each species.

Sampling (page 74)

1 Impossible to count all organisms in a habitat. It would also be too destructive.
2 To remove any bias on the part of the person doing the sampling. To obtain information that is representative of the habitat.
3 Two ways: (a) mark out a grid on the dunes; use random numbers to locate places to put quadrat; find percentage cover of marram grass (see figure on page 75); calculate a mean. (b) The line transect in the figure on page 74 shows that marram grass is not distributed evenly across the habitat, so quadrats could be put down along a line at regular intervals and the percentage cover recorded as in (a). This could be plotted on a graph to show how the abundance of marram grass changes across the dune.
4 Line transect is qualitative, but shows dominant vegetation at a glance; quadrats give quantitative data, but information is less easy to present. Estimating percentage cover and finding individual plants and animals (for species density) within a quadrat is often difficult.
5 Change in physical factors, e.g. high temperature, little rainfall; population explosion of one species; introduction of a new species, e.g. a predator; disturbance by humans.

Classification (page 76)

1 Classification: grouping organisms into categories according to similarities and differences between them.
Taxonomy: the science of classifying organisms.
Phylogeny: the evolutionary history of a group of organisms.
2 Natural classification systems like that shown on page 76 reflect the evolutionary history of organisms. Those that are closely related are grouped into the same category (e.g. genus) that share many common features. Those that are not closely related may be grouped into larger groups (e.g. phylum) with others that share a few fundamental features.
3 Organisms are classified into groups: the smallest group is the species. Similar species are classified into a genus; similar genera are classified into orders, and so on. Groups high in the hierarchy (e.g. phylum and class) contain organisms that share general features, for example, mammals suckle their young on milk and have hair and external ears (pinnae).
4 See the table on page 76. The taxa are in the first column beginning with domain and ending with species.
5 It will be given a binomial: generic and species names. The organism may be similar to other species, in which case there is no choice about the generic name. If unrelated to existing species, it will be given a new generic name. Suitable words are chosen – where it lives or who found it – and these are Latinised to give the scientific name.

The five kingdoms (page 78)

1 Unlike the organisms classified in the five kingdoms, viruses do not have a cellular structure.
2 Homologous features have the same pattern and share a common origin.
3 Answer could include description of several of the following: external appearance, anatomy, development, primary structure of proteins, fine detail of body surfaces using the SEM, DNA sequence data.
4 The three-domain classification: Archea, Bacteria, Eukaryota. Domain is above kingdom. Introduced because bacteria can be divided into two large groups according to features, such as ribosomal RNA, flagella structure and membrane structure.

Variation and adaptation (page 80)

1 Variation: differences between organisms, both between individuals in the same species and between species.
2 Between species – caused by differences in the genes; within species – caused by different alleles of the genes of that species; variation within a species is also due to the environment.
3 Discontinuous variation: different forms without any intermediates. Continuous variation: range of types with many intermediates; examples are features such as length and mass.
4 Adaptation refers to how a species 'fits in' to its environment. Organisms show adaptation at different levels: structural (morphological and anatomical), physiological, biochemical and behavioural. These contribute to the success of the organism so that it can survive, reproduce and pass on its alleles to the next generation. Adaptations evolve by natural selection and are constantly exposed to selection pressures. (The term is used in other contexts in biology, but this is the meaning you should know for this module.)
5 It is a good idea to choose examples of plants and animals that live in extreme environments, such as deserts and sand dunes (camels and marram grass). Microorganisms are also adapted to extreme environments, for example the extremophiles that live in habitats such as hot springs, where there are very high temperatures (see Examiner tip on page 56).

Natural selection (see page 82)

1 Darwin's four observations are listed on page 82. His arguments dealt with overproduction leading to competition for limited resources. The organisms that survive this 'struggle for existence' are better adapted to existing conditions and survive, breed and pass on their alleles to the next generation. (Darwin did not know about genes and alleles, but his argument is supported by our knowledge of genetics.)

2 Humans apply pesticides to kill insects, but the pesticide acts as a selective agent. Pests with mutations that provide protection against pesticides survive, breed and pass on their alleles.
3 Bacteria with mutations that provide protection against the antibiotics survive; other bacteria die or fail to reproduce. Resistant bacteria multiply and pass on the alleles that give protection, e.g. code for enzymes that break down antibiotics.

Conservation: maintaining biodiversity (page 84)

1 Ecosystems provide humans with goods, such as timber and fish, as well as services, such as sewage disposal and providing us with oxygen and removing carbon dioxide from the atmosphere. We should conserve ecosystems to ensure these goods and services are available in the future.
2 Keystone species: a species that plays a significant role in an ecosystem; examples are often carnivores that keep populations of grazers in check, e.g. sea otters and tigers.
3 Natural and man-made ecosystems (chalk downland, heathland, etc.) provide many goods and services. There are many complex interactions that occur within ecosystems (many that we do not fully understand). Losing species from ecosystems, or losing complete ecosystems, may have widespread effects, for example, removing a top carnivore results in a population explosion of grazers and a reduction in the vegetation they rely on. This could destroy the ecosystem, with the loss of many other species not directly linked with the top carnivore.
4 Agriculture: fertile lowlands are flooded by rising sea levels; many areas will become too dry to support crops. Some plant and animal diseases may spread to areas so far unaffected. Human health: vectors of tropical diseases will spread further north and south.

Conserving endangered species (page 86)

1 Their numbers are so small that they are at risk of extinction.
2 Some possible answers: species remain in their natural habitat where they are more likely to breed; many animals do not breed in captivity; only small numbers of large animals can be kept in captivity; animals lose ability to survive in wild after being kept in captivity.
3 Loss of habitat; dangers from poaching, hunting, illegal trade; numbers are too small to risk leaving in the wild.
4 Habitat destruction; presence of a predator or disease likely to wipe out a species; very small population with little genetic variability; breeding programmes in captivity can attempt to increase genetic variability by avoiding reproduction between animals that are very closely related.
5 As a signatory of the Convention on Biological Diversity, it had to!
6 Any area that is likely to be developed has to be surveyed by ecologists and the likely effects of the development on the area determined. These assessments are then used to inform the decisions made by planners, local and national governments before approval is granted.

Answers to end-of-unit questions

In these answers a semi-colon (;) separates individual marking points. Notice that in many cases there are more marking points available than there are marks for the question. This means that there are several ways in which you can gain full marks for a question. In some of the answers, you will find information that is not in the spreads – this is to give you extra help with the topics in the questions.

Unit 1 – Cells, exchange and transport

1 **a** X = vesicles; Y = ribosomes; Z = cell surface membrane (3)
b DBAC (1)
c Exocytosis (1)
d Golgi apparatus/Golgi body (1)

2 **a** It is made up of more than one type of tissue (1)
b A stem cell is unspecialised in form/structure; it is able to divide repeatedly/can divide continuously in culture; stem cells develop into specialised/differentiated cells; when they receive suitable chemical/hormonal signals; a differentiated cell is specialised to perform a function or functions in the body; it has a distinctive shape/structure/behaviour; some of its genes are active/switched on, others are inactive/switched off; many differentiated cells either never divide or show limited cell division (5)
c Red blood cells/erythrocytes; neutrophils/granulocytes; macrophages (2)
d Getting the cells to the site of damage may be difficult; they may develop into abnormal tissue/cancer cells/tumour cells; might not be available in emergency/after a heart attack; might be rejected/attacked by the patient's immune system; embryonic stem cells are produced by the destruction of a human embryo; this is thought (by some) to be unethical; although alternative method is removal of one cell from an 8-cell embryo so the embryo survives; signals/chemical stimuli needed to make stem cells specialise are not (fully) understood (3)
e Cambium cells/meristematic cells (1)

3 **a** A, external intercostal muscles; B, internal intercostal muscles (2)
b Atmospheric pressure (1)
c Elastic fibres/elastin/elastic walls (of thorax/alveoli) (1)
d During inspiration/breathing in, the diaphragm contracts/moves down, pushing the abdomen out; during expiration/breathing out, the diaphragm relaxes/returns to its domed position (2)

4 **a** They must breathe only from the (floating) box (1)
b People should not use the spirometer alone/unsupervised; the mouthpiece should be disinfected before use (with sodium hypochlorite solution); people with a respiratory infection/history of asthma should not use the spirometer; medical grade oxygen should be used to fill the box; if the box contains air, soda lime should not be included; dust-free soda lime must be used; care should be taken when filling/emptying the soda lime container; soda lime is corrosive/an irritant (3)
c Tidal volume is 500 – 600 cm^3/0.5 – 0.6 dm^3; vital capacity is 4200 cm^3/4.2 dm^3 (2)
d Oxygen is absorbed by the lungs/into the blood; so the volume in the box falls (steadily); carbon dioxide is not replacing oxygen in the box; because soda lime is absorbing carbon dioxide; gas/air/oxygen may be escaping from the apparatus if there is a leak (3)

5 **a** Any three from: nucleus/chloroplast/mitochondrion/ribosome/(rough or smooth) endoplasmic reticulum/Golgi apparatus/lysosome/vesicle/centriole (only in animal cells) (3)

Structures such as the nucleolus and cilia or undulipodia, that are not in the cytosol, are not valid answers to this question

b **i** Provides support/strengthens the cell; supports/holds the organelles; allows organelles to be moved about (inside the cell); allows the cell surface membrane to be moved/extended; allows the cytoplasm/cell to move or contract; moves chromosomes during nuclear division/mitosis/meiosis; brings about cytokinesis (3)
ii The cells are (very) large/long; axons/dendrites must be extended/supported; organelles/proteins/substances must be moved down the axon/to synapse (1)
iii The cell wall provides support; the cell does not contract/move; the (large) vacuole does not contain cytoplasm/organelles (1)

Unit 2 – Molecules, biodiversity, food and health

1 **a** Deoxyribose; phosphate (2)
b 29.2 because adenine pairs with thymine; also it will be the same as pig liver DNA (so could be 29.4) as it comes from the same animal; *note the comment about percentage error in the question* (2)
c There is no uracil in DNA, no thymine in RNA (1)
d Guanine and cytosine have the same percentage in each sample of DNA because they pair (in DNA but not in RNA); RNA is single-stranded; RNA produced/synthesised using only one of the two DNA strands/strand of the double helix; only some DNA sequences/genes used to produce RNA; some genes are switched off/not transcribed in each tissue (3)
e Root-tip cells are diploid, pollen haploid; pollen produced by meiosis (2)

2 **a** The volume of ONPG solution (in each tube) should be constant; the same concentration of ONPG would be used; temperature constant; by means of a water bath (preferably thermostatically-controlled); different buffer solutions used to make the pH vary; over a range of pH/e.g. pH 3 to 9; same volume of buffer solution would be used; the volume of enzyme solution must be controlled; the yellow colour of the product would be measured/assessed; after the same time; using a colorimeter; set the colorimeter to zero; record absorbance; use colorimeter to follow the reaction; *if no colorimeter* use a colour standard (yellow) and time how long it takes for reaction mixtures to reach the same colour; rate is 1 ÷ time in seconds (8)

b Hydrogen ions/pH changes, alter shape of enzyme molecule; ionic/hydrogen bonds break; shape of active site changes; substrate may not fit/bind as well; enzymes are proteins; pH alters tertiary structure (2)

c Galactose enters/binds to the active site; so fewer ONPG molecules can enter; enzyme–product complexes form (2)

3 a i (Common) salt/sodium ions/sodium chloride; refined carbohydrates rich in sugar/sucrose (2)

ii Fruit/vegetables; complex polysaccharides/plant fibres/ whole-grain cereals; antioxidants; omega-3 fatty acids (2)

b LDLs transport cholesterol in the blood; synthesised by the liver; deposited in the walls of (damaged) arteries; producing plaque/atheroma/atherosclerosis (4)

c There are many (risk) factors involved; any two from the following: genetic factors/stress/exercise; the French may experience protective factors that counter the harmful effects of fatty diet; e.g. antioxidants/fresh vegetables/red wine; (some) polyunsaturated fatty acids are protective (3)

4 a Habitat biodiversity is the variation shown by the physical environment and/or the types of community; e.g. wetter and dryer areas/older and more recently forested areas; species biodiversity is the number and abundance of different species; planners should avoid routes where biodiversity is high (4)

b Genetic diversity is the variation in the DNA/genes of a species; high genetic diversity is where there are several alleles for (most) of the genes in a species (2)

c Choosing random quadrats using map/grid/random coordinates is unbiased; makes sure places that are difficult to reach are fairly represented; selection by an experienced observer makes it less likely that important/interesting sites are not left out by chance; the full range of the habitat diversity is recorded/noted; experienced observer may miss (important) detail (from a helicopter); random sample requires more quadrats to avoid the effects of chance (3)

d i (n/N) for red-breasted flycatcher = 0.211; Simpson's diversity index is 0.182 (2)

ii Species richness is the number of different species (of bird) (1)

e Only one quadrat, so not a significant sample; counting at fixed time each day; may miss some species/bias in favour of some species; counting only during one week/one time of year; some birds are territorial/will exclude others (which are therefore not counted); same individuals may be counted repeatedly (4)

f Natural forest may have more food; nest sites; trees may be older; more variable in size/age; greater diversity of plants/ insects/species that birds eat; *Eucalyptus* plantation uniform/ lower habitat diversity; specific insects feed on *Eucalyptus* in Australia (but not in Portugal) (3)

Index